Whose Science? Whose Knowledge?

Whose Science?
Whose Knowledge?

Thinking from Women's Lives

Sandra Harding

Cornell University Press

Ithaca, New York

First published 1991 by Cornell University Press.
First printing, Cornell Paperbacks, 1991.
Second printing 1992.

Library of Congress Cataloging-in-Publication Data

Harding, Sandra G.
    Whose science? whose knowledge? : thinking from women's lives / Sandra Harding.
        p. cm.
    Includes bibliographical references and index.
    ISBN 0-8014-2513-1 (cloth : alkaline paper). — ISBN 0-8014-9746-9 (paper : alkaline paper)
    1. Women in science. 2. Feminist theory. 3. Knowledge, Theory of. 4. Science—Social aspects. I. Title.
Q130.H37 1991
305.43'5—dc20   90-55724

Printed in the United States of America

⊗ The paper in this book
meets the minimum requirements of the American
National Standard for Information Sciences—
Permanence of Paper for Printed Library Materials,
ANSI Z39.48-1984.

Contents

Preface

In *The Science Question in Feminism* (1986) I showed how feminist criticisms raised important issues not only about the social structure and uses of the sciences but also about the origins, problematics, social meanings, agendas, and theories of scientific knowledge-seeking. Feminists were asking "the woman question" in science: "What do women want from the sciences and their technologies?" But they were also asking "the science question in feminism": "Is it possible to use for liberatory ends sciences that are apparently so intimately involved in Western, bourgeois, and masculine projects?" One theme in that book was the importance of tensions between three epistemological programs: feminist empiricist philosophy, which tries to correct "bad science"; feminist standpoint theory, which tries to construct knowledge from the perspective of women's lives; and feminist postmodernism, which is suspicious of the Enlightenment loyalties inherent in such scientific and epistemological projects.

Several themes of that book are pursued further here. Because the social and intellectual contexts for thinking about women, feminism, science, and knowledge have been shifting rapidly, however, certain projects have moved in somewhat different directions from those which that study anticipated. For one thing, I now see important ways to develop the intellectually powerful feminist standpoint theory of knowledge to meet a good number of questions I raised about it in the earlier book. This epistemology, like the feminist research and scholarship for which it provides a metatheory, can be pried yet further away from the excessively Western and "modernist" constraints of the older theory from which it fruitfully borrowed.

In the last five years there has been an outpouring of critical examinations of Western science, technology, and epistemology from the peace and ecology movements, the left, philosophers, historians, and sociologists of science, and Third World critics, as well as from Western feminists. Consequently, the feminist critiques are not isolated voices crying in the wilderness (if they ever were) but are linked thematically and historically to a rising tide of critical analysis of the mental life and social relations of the modern, androcentric, imperial, bourgeois West, including its sciences and notions of knowledge. The present book is a study about this moment in the feminist criticisms of science, technology, and epistemology. It is *a* study, not *the* study; other participants in these debates would focus on issues other than the ones I have chosen. It is a book about Western sciences, technologies, and claims to knowledge from feminist perspectives, and in the later chapters it begins to be about appropriate focuses on such issues from the perspective of "global feminisms"—feminisms capable of speaking out of particular historical concerns other than the local Western ones that continue to distort so much of Western intellectual life.

I have not been concerned to report on or to argue with other philosophers, science observers, or feminists except where doing so seemed to advance the arguments of this book. My thought has, however, been informed especially by work being done in three contemporary intellectual areas that are themselves overlapping, interwoven, and occasionally in tension with one another. First and foremost are the writings of other feminists on science and epistemology and on political philosophy in all its leading contemporary forms. African American, Third World, socialist, and postmodern focuses in feminist political philosophy have proved particularly helpful. Second, critiques of Eurocentrism by African and African American philosophers, African American feminists, and Third World writers have brought into focus the importance of considering on whose questions the Western sciences, philosophy, and social studies of the sciences (including feminist ones) have centered, and the necessity of historicizing "science," "women," and "feminism." And third, I have benefited from the work of contemporary philosophers, sociologists, and historians writing in the social studies of science.

This book not only profits from these intellectual traditions; it is also addressed to activists and other thinkers within them. Consequently, it is intended to reach some readers who are just beginning to consider the contributions they could make to a more realistic understanding of the relations between feminism, science, and social life. Not every chapter is addressed to all these audiences. By the later chapters, however, I hope to have created a context that empowers readers to begin to think simultaneously both within and about all these contemporary intellectual and political currents.

In a sense there are two introductions to the book. The formal one in Chapter 1, "After the Science Question in Feminism," explores the historical context within which contemporary feminist thought about science and knowledge occurs, and identifies five issues that are examined in the following chapters; the opening chapter of Part I, "Feminism Confronts the Sciences: Reform and Transformation," surveys major themes in the feminist science literature at this moment and suggests the relevance of these critiques to contemporary intellectual and political concerns.

While each of the other ten chapters is designed to be read on its own, two sustained arguments are pursued throughout. First, we need a more complex understanding of how the development of Western sciences and models of knowledge are embedded in and have advanced the development of Western society and culture but have also led to the simultaneous de-development and continual re-creation of "others"— Third world peoples, women, the poor, nature. Science and knowledge will always be deeply permeated by the social relations through which they come into existence, but it is contemporary social relations that create and recreate science and knowledge today. The sciences that gain world ascendancy in the future are unlikely to be so distinctively Western and androcentric as are the dominant tendencies today. Conventional thinkers have argued, on the one hand, that the natural sciences must be the model for critical social sciences and, on the other hand, that the natural and social sciences are more or less irrelevant to each other. But if natural sciences and their preoccupations in reporting on nature are embedded in and complicitous with social projects, then a causal, scientific grasp of nature and how to study it must be embedded in—be a special area of—causal, scientific studies of social relations and how to study them.

Second, feminisms must become capable of influencing other liberatory movements as they are simultaneously influenced by them. Many observers have noted the difficult relationships between various progressive political movements and their science and knowledge agendas. For example, feminists concerned with the role of class relations in structuring social life or with ecological issues are often repelled by the not so subtle sexism and androcentrism that permeates much of the left and the ecology movement. There is no doubt, however, that many people concerned about class and race issues have found certain aspects of feminism part of the problem that *their* work seeks to resolve. If feminisms are to succeed, they cannot promote a binary opposition between social groups named "women" and "men," though they must still remain firmly feminist. Increasingly replacing the focus on male supremacy that preoccupied much feminist writing of the 1970s are new analyses of gender relations as they have historically been constructed through imperialism, class exploitation, and the control of sexuality. These studies are not intended to allay men's fears that "women's studies" or "feminist studies" are too political; they are fully political, as every account produced by humans cannot help being. But they take a broader and more reflexive field for their own analyses: they historicize, contextualize within history, the male supremacy that has been the particularly prevalent and insufferable part of gender relations.

Part I of the book, "Science," focuses on issues of special interest to natural scientists and researchers working in the social studies of the natural sciences. It raises some new critical issues about the sciences and is designed to make the later discussions of these topics more accessible to readers who are new to feminist critiques of science and epistemology. Part II, "Epistemology," pursues some of the issues that the "standpoint theories" of knowledge raise about traditional epistemologies, critically examines the assumptions and logical consequences of the theories, and compares this kind of epistemology with postmodern theories of knowledge. Part III, "Others," pursues the logic of taking standpoint epistemologies as a directive to begin our thinking from the standpoint of the lives of groups that have not been central to Western feminist discussions of science and epistemology (let alone to the dominant discourses). Then it reconsiders the relationship between experience and knowledge and asks what the liberatory movements can do to hasten the "birthing" of new agents of history and knowledge. The concluding chapter argues that within the transformed logics of feminism and of science which have been the subject of the preceding chapters, it makes sense to think that distinctively feminist sciences have already appeared.

According to the dramatic theory of playwright Bertolt Brecht, the structure of the Aristotelian drama should be rejected as detrimental to progressive politics. The Aristotelian tragedy provided a catharsis for its audiences that was all too comforting. Theater for progressive social change should not encourage the spectators to pour out into the streets delighted to be alive and well and ready to resume their daily lives after their "bad dream" in the theater; such a narrative structure dampens and suppresses whatever tendencies toward progressive social change the audience might have, Brecht thought. Feminist science and epistemology cannot afford a simplistic and satisfying Aristotelian plot; they must follow a more disorienting Brechtian story line. To attempt to resolve within one book the tensions, antagonisms, and contradictions in the encounters between feminism and Western models of science and knowledge would be not only foolish, but intellectually and politically regressive. Instead, I hope to advance more useful ways for us to think about and plan their future encounters. The truth (whatever that is!) cannot set us free. But less partial and less distorted beliefs—less false beliefs—are a crucial resource for understanding ourselves and others, and for designing liberatory social relations.

More people than I can name have contributed to shaping this book through their helpful comments on earlier drafts and their questions in response to lectures on these themes. For critical readings of the manuscript or parts of it, I am grateful especially to Christien Brouwer, Judith Butler, Anne Fausto-Sterling, Evelynn Hammonds, Frances Hanckel, Nancy Hartsock, Ruth Hubbard, Alison Jaggar, Anna Jonasdottir, Margareta Lindholm, and Judith Roof. For conversations through the years that have provoked me to change the way I think, I am indebted particularly to Margaret Andersen, Cynthia Enloe, Jane Flax, Donna Haraway, and Nancy Hartsock.

A fellowship at the University of Delaware Center for Advanced Studies in 1989–90 gave me the time to revise Chapters 6, 8, 9, and 12. The support and good advice of Frank Dilley at the University of Delaware and John G. Ackerman at Cornell University Press have been invaluable resources. Bill Lawson came up with the winning title. Patricia Sterling's copyediting made this a much better book. Frances Hanckel's enthusiasm and support have made the struggle of producing a book much more fun.

Chapter 3 is a slightly revised version of a paper that appeared in *Women's Studies International Forum* 12, no. 3 (1989), 271–283 (© Pergamon Journals Limited). An earlier version of the discussion of feminist standpoint theory in Chapter 5 appeared in the *Journal of Social Philosophy* 21 (1990).

SANDRA HARDING

Newark, Delaware

Whose Science? Whose Knowledge?

I

Introduction

After the Science Question in Feminism

The feminist discussions of science, technology, and theories of knowledge occur at a moment of rising skepticism about the benefits that the sciences and their technologies can bring to society. Calls for reforms and transformations have arisen from many different groups. However, these discussions also occur when intellectuals in the fields of science and technology are gaining more and more power in higher education and in government.

Feminists themselves are of at least three minds about the sciences. They (we) criticize not only "bad science" but also the problematics, agendas, ethics, consequences, and status of what has come to be called "science-as-usual." The criticisms of science-as-usual are made in the context of a call for better science: important tendencies within feminism propose to provide empirically more adequate and theoretically less partial and distorted descriptions and explanations of women, men, gender relations, and the rest of the social and natural worlds, including how the sciences did, do, and could function. From theorists who draw on European philosophy, however, comes criticism of the very idea of trying to reconstruct science, whether or not in feminist ways. These feminists appear to be arguing that there is no baby to be found in the bath water we would throw out. Additionally, analyses flow from not just one but many feminisms, each increasingly well developed in both theoretical and historical terms. Consequently, feminist analysts of science, technology, and epistemology disagree with one another over many important aspects of these issues.

Feminism and Science: A Confusing Moment

Skepticism about the Sciences

Modern Western sciences and their technologies have always been regarded with both enthusiasm and dread. On the one hand, we tend to attribute to them at least some responsibility for the high standards of living that many in the West enjoy—especially if we are white and middle or upper class. It is unimaginable to us that we could want to give up the food and clothing, medical treatment, cars and airplanes, computers, television sets, and telephones that have become available through scientific and technological development. On the other hand, just who or what is responsible for atomic bombs, Agent Orange, industrial exploitation, polluted air and vast oil spills, dangerous contraceptives such as Dalkon shields, inappropriate uses of Valium, health profiteering, high infant mortality in the United States, famine in Ethiopia, and the development of a black underclass in the United States? Conventionalists insist that science get full credit for the good aspects of the "Western way of life" but that such "misuses and abuses" are entirely the fault either of politicians or of the industries that apply supposedly pure information in socially irresponsible ways.

The insistence on this separation between the work of pure scientific inquiry and the work of technology and applied science has long been recognized as one important strategy in the attempt of Western elites to avoid taking responsibility for the origins and consequences of the sciences and their technologies or for the interests, desires, and values they promote. From a sociological perspective, it is virtually irresistible to regard contemporary science as fundamentally a social problem. Sal Restivo has argued that it should be conceptualized as no different in this respect from alcoholism, crime, excessive drug use, and poverty.[1] The name "Frankenstein," which Mary Shelley gave to the scientist in her dystopian novel, has in popular thought migrated to the monster he inadvertently created. How the monster actually got created—and gets nourished and reproduced day after day—retreats into the shadows, as if there are no persons or institutional practices that we can

---

1. Sal Restivo, "Modern Science as a Social Problem," *Social Problems* 35:3 (1988).

hold responsible for the shape of the sciences and the kind of social order with which they have been in partnership.

These kinds of issues have been raised by feminists (see Chapter 2), but they are certainly not what is unique about feminist analyses. In one form or another, such concerns are voiced by the ecology and environmental movement, the peace movement, the animals rights movement, leftist and worker movements, and antiracist and antiimperialist movements in both the West and the Third World. Even "postmodernist" criticisms of the philosophical foundations of Western rationality should be regarded as part of the counterculture of science. What is at issue for all these critics, including feminists, is not only the easily identifiable theories, methods, institutions, and technological consequences of the sciences but also something harder to describe: the Western scientific world view or mind-set. The "indigenous peoples" of the modern West—those most at home in Western societies—have culturally distinctive belief patterns in which scientific rationality plays a central role. These "natives," like all others, have trouble even recognizing that they exhibit culturally distinctive patterns of belief; it is like discovering that one speaks a distinctive genre—prose. From an anthropological perspective, faith in scientific rationality is at least partly responsible for many of the Western beliefs and behaviors that appear most irrational to people whose life patterns and projects do not so easily fit with those of the modern West. From the perspective of women's lives, scientific rationality frequently appears irrational.

Still, scientific rationality certainly is not as monolithic or determinist as many think or as the description above suggests. Nor is it all "bad." It has been versatile and flexible enough throughout its history to permit constant reinterpretation of what should count as legitimate objects and processes of scientific research; it is itself shaped by cultural transformations and must struggle within them; and it is inherently no better or worse than other widespread social assumptions that have appealed to groups with different and sometimes conflicting agendas. Perhaps even liberalism and feminism would provide examples, since both have at times been associated with racist and bourgeois projects, even though at other times they have advanced struggles against racism and class exploitation. It is one theme of this book that modern Western science contains both progressive and regressive tendencies, and that our task must be to advance the former and block the latter. Indeed, scientific rationality can make possible the transformation of its own agendas; critics from feminist and other scientific countercultures certainly intend to use it for this purpose.

The Rising Status of the Intellectuals in Science and Technology

Increased participation in the countercultures of the sciences occurs just when the prestige of the intellectuals in science and technology is rising in higher education and the government. Scientists have been held in high regard since Sputnik, of course—indeed, even since Newton[2]—but the flood of industrial and federal funds that pours into scientific and technological projects in universities these days is truly astounding. It is a long time since scientific research could be regarded as significantly isolated in real life from the goals of the state and industry—if it ever could. Scientific research is an important part of the economic base of modern Western societies.

No doubt envy plays a certain role in the criticism of science. Scholars from the humanities and social sciences perceive themselves increasingly working in offices cramped into university attics and basements as new science and engineering buildings open; they lose what they think are too many of the best graduate students to the sciences and engineering as they lose support for graduate programs. More and more they find themselves reporting to deans, provosts, presidents, chancellors, and trustees whose backgrounds are in science and engineering and who intend to take universities where the money is flowing. How could they justify not doing so, these administrators ask.

Intellectuals in science and technology do not see their situation as rosy. One can hardly open a science journal or even an airline magazine without finding hand-wringing projections of a shortage of scientists and engineers. It has gotten so bad, they say, that in order to "keep America strong" they are even willing to develop special programs to recruit women and minorities to science, mathematics, and engineering departments. This institutional setting needs to be kept in mind when one thinks about the "postmodernist" criticisms of the philo-

2. See, e.g., Wolfgang Van den Daele, "The Social Construction of Science," in *The Social Production of Scientific Knowledge*, ed. Everett Mendelsohn, Peter Weingart, and Richard Whitley (Dordrecht: Reidel, 1977).

sophical foundations of modern science. The attractions of the postmodernist critique are many,[3] but among them are surely its perceived usefulness as a means to restore status to the humanities, status that has stolen away to science and technology without public discussion of the benefits and losses of such a move.[4] The intellectual fundamentalism of Allan Bloom and the "back to the classics" movement in the United States is another critical response to the rise in status of science and technology. The countercultures of science have at least the beginnings of a realistic assessment of possible futures for the West, an assessment that is lacking in intellectual fundamentalism.

The Need for New Sciences

It is at this moment that feminism and other liberatory social movements appear on the scene with agendas that include generating new sciences. Women need sciences and technologies that are *for* women and that are for women in *every class, race, and culture*. Feminists (male and female) want to close the gender gap in scientific and technological literacy, to invent modes of thought and learn the existing techniques and skills that will enable women to get more control over the conditions of their lives. Such sciences can and must benefit men, too—especially those marginalized by racism, imperialism, and class exploitation; the new sciences are not to be *only* for women. But it is time to ask what sciences would look like that were *for* "female men," all of them, and not primarily for the white, Western, and economically advantaged "male men" toward whom benefit from the sciences has disproportionately tended to flow. Moreover, it is time to examine critically the conflicting interests in science that women in opposing classes and races may well have; women's interests are not homogeneous. Feminism insists that questions be asked of nature, of social relations, and of the sciences different from those that "prefeminists" have asked, whether conventional or countercultural. How can women manage their lives in the context of sciences and technologies designed and directed by powerful institutions that appear to have few interests

3. And so are the problems with it, many will say. Postmodernism is discussed in later chapters.
4. Philosopher Cornel West made this point in the plenary session "What Is Cultural Studies?" at the conference sponsored by the Committee for Cultural Studies, City University of New York, May 11, 1989.

in creating social relations beneficial to anyone but those in the dominant groups?

Thus, though it would be foolish to deny that science is a major social problem, we can ask who benefits from regarding it as *nothing but* a social problem. Possible scientific beliefs and practices are not limited to those that have already existed, let alone to that subset that has existed in the modern West. It is complicitous with the dominant ideology to assert that everything deserving the name of science has been done in the modern West. Nevertheless, we must contend realistically with what the West has done with its sciences. It is important for the countercultures to struggle with science and technology on the existing social terrains while they also try to envision and plan different social environments for science in the future.

The Diversity of Feminist Analyses

Feminist analyses of science, technology, and knowledge are not monolithic. There is no single set of claims beyond a few generalities that could be called "feminism" without controversy among feminists. (The same could be said about sexism or androcentrism or non-feminism, which can also claim diverse historical frameworks and projects: Aristotle is not Freud.) The feminist science discussions are both enriched and constrained by the different political, practical, and conceptual perspectives that they bring to bear on science, its beliefs, practices, and institutions.

This is a good place to note that the term "feminism" is itself a contested zone not only within feminism but also between feminism and its critics. It is widely used as a critical epithet in the Second and Third Worlds and in some Western subcultures, by women as well as by men, to prevent women from organizing across class, race, and national borders and even just to "keep women in their place."[5] It is also important to note that widespread tendency in the West, at least, for women and men to insist that they are absolutely not feminists but then to advance the very same intellectual and political programs that are promoted by others under the label of feminism. These non-feminists too are for ending violence against women, the sexual exploi-

5. The designations First, Second, and Third Worlds have been constructed by the West. They distort global politics in many ways, all to the benefit of the West, but I use them for lack of better terms.

tation of women, women's poverty, job discrimination against women, the exclusion of women from public office, unequal educational opportunities, sexist biological and sociological and historical claims, and so on. For these people, "feminism" appears a handy label for those elements in feminism from which they wish to distance themselves—and it is the Eurocentric, racist, bourgeois, and heterosexist elements in feminism, as well as the vigorous opposition to them, from which different groups wish to distance themselves.

I think it is important to try to distinguish regressive from progressive tendencies in peoples' actions and beliefs and to support the progressive tendencies, whether or not others think about them in just the way I do. What appears to be radical and progressive from the perspective of some women's lives may be too conservative, too dangerous, or just irrelevant from the perspective of other women's lives. If feminism is a term people find appropriate to their attempts to improve women's conditions, they will use it. It would be regressive and ethnocentric for me to decide for them that they should adopt a term I find useful in my world. Nevertheless, I do use the word throughout this book, since I can assume that the majority of readers will find it appropriate here.

Several distinctive traditions of thought within which feminists have analyzed human nature, the fundamental causes of women's inferior conditions, and what should be done to change those conditions generate different issues about science, technology, and epistemology. Most important are the "grand theory" traditions that borrow from Western political theory: liberal feminism and traditional Marxist feminism. We should also include in this group the African American feminism that has strong roots, we are now learning, in the nineteenth-century struggles of African American women.[6] Then there are the now well-developed feminisms that emerged in the politics of the 1960s: radical feminism, socialist feminism, and the feminisms of racially marginalized women both in the West and in the Third World, some associated with national liberation struggles.[7] Other feminist political orientations and traditions can be located within and along-

6. See, e.g., Hazel Carby, *Reconstructing Womanhood* (New York: Oxford University Press, 1987); Angela Davis, *Women, Race, and Class* (New York: Random House, 1981); Paula Giddings, *When and Where I Enter: The Impact of Black Women on Race and Sex in America* (New York: Bantam Books, 1985).

7. See Alison Jaggar, *Feminist Politics and Human Nature* (Totowa, N.J.: Rowman & Allenheld, 1983).

side these: anarchist feminism, Jewish feminism, lesbian and gay feminisms, antimilitarist feminism, ecology-focused feminism, and others. Most of these feminists also work in other intellectual and political movements, as their compound identities indicate. Each of these "movement" orientations brings unique concerns and approaches to discussions of gender, science, and knowledge.

Moreover, feminists work in diverse social settings. In the United States we work in battered women's shelters and rape crisis centers, in agencies for international development and mainstream political organizations, in law and medicine, in child-care and organizational management, in factories and secretarial pools, in computer programming and therapy—not to mention in laboratories and women's studies programs. And we experience the consequences of developments in science and technology not only at work but also as pregnant women and mothers, as sick or old, as pedestrians or drivers, and every time we eat or even breathe. We experience science and technology in our everyday lives, in the struggles for dignity and survival that women engage in daily on behalf of their kin and community as well as themselves.[8] In Western Europe and the Second and Third Worlds, there are other culturally specific daily activities of women that produce distinctive experiences of Western science and technology. It is in different and conflicting ways that women experience modern science and technology in each of these locations. Analyses from these different social perspectives have contributed insights—sometimes contradictory ones—to our understanding of the sciences and their technologies.

Additionally, the conceptual frameworks and current agendas of our disciplines and the various approaches within them have provided important resources for feminist science discussions. Feminist analyses have drawn from the history of science, focused on intellectual or social history, formal and informal institutions, economic history, or the history of individuals; from the sociology of science, focused on the structure of occupations, the workings of institutions, the legitimation of erroneous belief, the class structure of science, the sociology of knowledge, or the microstructure of laboratory life; from the philosophy of science, informed by traditional rationalist and empiricist agendas, Marxist epistemology, critical theory, the postmodernism of Jean-

8. See Bettina Aptheker, *Tapestries of Life: Women's Work, Women's Consciousness, and the Meaning of Daily Life* (Amherst: University of Massachusetts Press, 1989).

François Lyotard, Michel Foucault, Richard Rorty. All these theoretical and disciplinary frameworks—and others, such as literary criticism, psychoanalysis, and even art history—have provided rich resources for the study of gender and science.[9] At the same time, the "prefeminist" schemes have limited or obscured important ways in which the relations between women, gender, and science could and should be analyzed.

A Complex and Changing Environment for Discussion

The joint action of these various competing and interacting forces in the terrain in which feminism also operates—indeed, feminism is also part of all of these other tendencies—will have consequences different from those one might imagine from the perspective of the feminist critiques alone. It is as if we were at the point at which bands of men and women leave the familiar streets of their different neighborhoods to join an ongoing march down a boulevard. We watch each band struggle to maintain its identity and carry its banners forward as it is jostled by boisterous groups with similar intent. As the crowd surges forward, some people leave their group to join others; some groups merge, and others disappear. The words of anthems change, and the inadvertent harmonies and disharmonies created when one hears two bands playing at once suggest previously unimaginable musical possibilities—not all of them desirable. The necessity to struggle to advance their goals in the environment of everyone else's equally determined efforts creates configurations different from those of individual groups marching alone. Similarly, feminist tendencies must struggle against, with, and within these other streams of contemporary intellectual, political, and social life. The consequences of these interactions cannot but be surprising to everyone.

Challenges

Five issues that are at present emerging in one form or another from recent analyses of science, technology, epistemology, and feminism shape my concerns in the chapters that follow. The challenge in each

9. One good place for newcomers to start in this literature is the collection of essays in Sandra Harding and Jean O'Barr, eds., *Sex and Scientific Inquiry* (Chicago: University of Chicago Press, 1987). See also the sources cited in Chapter 2.

case is to develop conceptual frameworks that are theoretically rich enough and empirically adequate to enable us to think what appear at first to be contradictory thoughts.

(1) Science is politics by other means, and it also generates reliable information about the empirical world. Science is more than politics, of course, but it is that. It is a contested terrain and has been so from its origins. Groups with conflicting social agendas have struggled to gain control of the social resources that the sciences—their "information," their technologies, and their prestige—can provide. For those who have suffered from what seem to be the consequences of the sciences, their technologies, and their forms of rationality, it appears absurd to regard science as the value-free, disinterested, impartial, Archimedean *arbiter* of conflicting agendas, as conventional mythology holds.

And yet sciences created through political struggles, which are the only ones we have ever had, usually do produce reliable information about nature and social relations—reliable, that is, for some group or another's purposes. They are no less sciences for being driven by particular historical and political projects.

There are few resources in the conventional philosophy of science, epistemology, or sociology of science, however, which permit the articulation and exploration of these seemingly contradictory understandings. It is a challenge for feminism and other countercultures of science to develop conceptual frameworks that encourage widespread discussion of this apparently contradictory character of science.

(2) Science contains both progressive and regressive tendencies. So does feminism. To say this about science is to oppose the view that "science is inherently good, although it is sometimes applied in regressive ways." And it is to oppose the view that "science is inherently value-neutral, although it can be used in progressive or regressive ways." It is to oppose both views because they refuse to recognize that the social origins of science and the values it carries suffuse scientific projects. A critical examination of these origins and values can be carried out as part of the project of science, however. The very scientific rationality that has been the object of criticism from so many quarters contains the resources for its own transformation. Thus, what science becomes in any historical era depends upon what we make of it.

The same can be said of feminism. It too contains both progressive and regressive tendencies. It is not usefully conceptualized without qualification as inherently good—and of course no one characterizes it as value-neutral—because its origins and the values it carries clearly shape its projects. Those of its tendencies that focus on male supremacy and gender relations without giving equal weight to other important aspects of social relations can provide resources for Eurocentrism, racism, imperialism, compulsory heterosexism, and class exploitative beliefs and practices—whether or not such a result is overtly or consciously intended. But it also contains tendencies that can contribute sturdy resources to the elimination of these forms of oppression, exploitation, and domination.

It is a challenge for feminism and other contemporary countercultures of science to figure out just which are the regressive and which the progressive tendencies brought into play in any particular scientific or feminist project, and how to advance the progressive and inhibit the regressive ones. The countercultures of science must elicit and address these contradictory elements in the sciences and scientific consciousness (and feminists must continue to do so with their various feminisms). The alternative is that regressive forces in the larger society manipulate these contradictory features and mobilize the progressive tendencies for their own ends. For example, international financiers appeal to belief in scientific and technological progress to gain support for technology transfers to the Third World which deteriorate the power of people there to control their lives. In the West it appears that there must be something wrong with "those people" if they cannot progress even when "gifted" with the supposed fruits of First World science and technology. Industries appeal to feminist themes about the importance of new health standards for women in order to produce profit from the sales of sporting goods, cosmetics, and so-called "health food."

(3) The observer and the observed are in the same causal scientific plane. An outpouring of recent studies in every area of the social studies of the sciences forces the recognition that all scientific knowledge is always, in every respect, socially situated.[10] Neither knowers nor the knowledge they produce are or could be impartial, disinterested, value-neutral, Archimedean. The challenge is to articulate how it is that knowledge has a socially situated character denied to it by the conventional view, and to work through the transformations

---

10. Donna Haraway focuses on the importance of his insight and supplies the useful term: "Situated Knowledges: *The Science Question in Feminism* and the Privilege of Partial Perspective," *Feminist Studies* 14:3 (1988).

that this conception of knowledge requires of conventional notions such as objectivity, relativism, rationality, and reflexivity.

Another way to put the issue is to note that if science is created only within political struggles, as mentioned above, then our "best beliefs," not just the least defensible ones, have social causes.[11] This means that observers and their subject matters are in the same social, political, economic, and psychological scientific planes. If, as the social sciences hold, class and race and gender relations must be called on to explain observable patterns in the social beliefs and behaviors of other people—of health profiteers, or the Ku Klux Klan, or rapists—then other aspects of those very kinds of relations have probably shaped the "empirically supported," "confirmed by evidence," and therefore less false results of our own fine research projects as well. We should think of the social location of our own research—the place in race, gender, and class relations from which it originates and from which it receives its empirical support—as part of the implicit or explicit evidence for our best claims as well as our worst ones.

One consequence of this claim is that we can understand how inanimate nature *simulates* encultured humans in that it always comes to us culturally preconstructed as a possible object of knowledge, just as do humans. Humans construct themselves as possible objects of knowledge and have also constructed inanimate nature as a possible object of knowledge. We cannot "strip nature bare" to "reveal her secrets," as conventional views have held, for no matter how long the striptease continues or how rigorous its choreography, we will always find under each "veil" only nature-as-conceptualized-within-cultural projects; we will always (but not only) find more veils. Moreover, the very attempt to strip nature bare weaves more veils, it turns out. Nature-as-an-object-of-knowledge simulates culture, and science is part of the cultural activity that continually produces nature-as-an-object-of-knowledge in culturally specific forms.

Neither the conventional nor the countercultural science discussions have developed conceptually rich enough or empirically adequate frameworks to enable critical thought about the fact or consequences of recognizing that observers and observed are in the same scientific

---

11. The last part of this claim is the contention of the "strong programme" in the sociology of knowledge (to be discussed later in this book), with which I agree in this respect, though not in others. See, e.g., David Bloor, *Knowledge and Social Imagery* (London: Routledge & Kegan Paul, 1977).

field. This understanding brings into sight a new kind of agent of both knowledge and history.

(4) It is necessary to decenter white, middle-class, heterosexual, Western women in Western feminist thought and yet still generate feminist analyses from the perspective of women's lives. Feminists have argued for the decentering of masculinity in society's thoughts and practices: no longer should manliness (however that is culturally defined) be the standard for the so-called human; no longer should masculinity and its widespread expressions across the canvas of cultural life be the preoccupation of everyone's anxious attention. The centering of men's needs, interests, desires, visions ensures only partial and distorted understandings and social practices. (And it *must* be possible for women to criticize this institutionalization of masculinity without being thought to "hate men.")

But then it is also necessary to decenter the preoccupations of white, economically advantaged, heterosexual, and Western feminists in the thinking and politics of feminists with these characteristics. No longer should their needs, interests, desires, and visions be permitted to set the standard for feminist visions of the human or to enjoy so much attention in feminist writings. How can this decentering be enacted in the discussions and practices of feminist science and technology? What will be feminist about them if they are not grounded in the presumed common lives of women?

One way to approach this issue is to keep in mind the argument of Jane Flax and others that gender is fundamentally a relation, not a thing.[12] That is, masculine and feminine are always defined "against each other," though the "content" of womanliness and manliness can vary immensely. Furthermore, as Judith Butler argues, gender is not an "interior state" but a performance that each of us acts and reenacts daily.[13] Moreover, we can see that the relationship picked out by "woman" or "man" is always a historically situated one. It is not constructed by relations between men and women in general, for there are no such persons and therefore no such relations. Nor are the gender relations between men and women in any particular group shaped only by the men and women in that group, for those relations too are

12. See Jane Flax, *Thinking Fragments: Psychoanalysis, Feminism, and Postmodernism in the Contemporary West* (Berkeley: University of California Press, 1990).

13. Judith Butler, *Gender Trouble: Feminism and the Subversion of Identity* (New York: Routledge, 1990).

always shaped by how men and women are defined in every other race, class, or culture in the environment. Gender relations in any particular historical situation are always constructed by the entire array of hierarchical social relations in which "woman" or "man" participates. The femininity prescribed for the plantation owner's wife was exactly what was forbidden for the black slave woman.[14] The forms of femininity required of Aryan women in Nazi Germany were exactly what was forbidden—and in fact eliminated—for women who were Jews, Gypsies, or members of other "inferior races."[15] So we cannot meaningfully talk about "women and science" or "women and knowledge" without exploring the different meanings and practices that accumulate in the life of someone who is a woman at any particular historical intersection of race, class, and culture. There are as many relationships between women and science as there are cultural configurations of womanhood (and of science).

Being white or Western or economically advantaged or heterosexual, however, need not be the scientific and epistemological disadvantage that one might expect it to be when one thinks about these identities as parallel to *andocentric* ways of being a man in gender relations. To decenter manliness does not mean that men can make no contributions to feminism or can generate no original feminist insights out of their own particular historical experiences. At least some have already done so. Similarly, white women can (and do) generate original antiracist insights out of their particular historical experiences as white women. We can demand of ourselves that we do so as a condition of producing analyses and politics adequate to feminism in a global context. But just what we are to demand of ourselves from such apparently contradictory social situations as "male feminist" and "white antiracist" requires more analysis than it has yet received.[16]

(5) The natural sciences are illuminatingly conceptualized as part of the social sciences. What kind of theoretical framework will enable us to understand sciences-in-society and the consequent society-in-sciences? According to one influential tendency in conventional thought,

14. Davis, *Women, Race, and Class*.

15. Gisela Boch, "Racism and Sexism in Nazi Germany: Motherhood, Compulsory Sterilization, and the State," *Signs* 8:3 (1983).

16. Questions have been raised (by me, among others) about the ability of the feminist standpoint epistemology to deal with differences between women's lives. Here, however, I defend the theory against these and related skeptical questions.

there is only one standard for what counts as science, and that is provided by the natural sciences. Physics, with its reliance on quantitative methods and its positivist ethos, is supposed to be accorded the highest rank among the natural sciences, with chemistry and then the more abstract areas of biology following behind. The social sciences are even lower on this scale. The "harder" social sciences such as economics and behaviorist psychology (cognitive psychology would now probably be substituted for behaviorism) lead the "softer" fields (softer to the extent that they rely more on "qualitative" studies) such as anthropology, sociology, and history. Some writers have even thought that the natural sciences should be the model for all knowledge, certainly for anything deserving such prestigious words as "scientific," "rational," and "objective." The sciences are fundamentally one, and the model for that one is physics. This internal ordering reflects fairly accurately the power and prestige accorded different fields of research within the sciences today.

Such a conception, however, prevents us from developing *natural* as well as social sciences that are not systematically blinded to the ways in which their descriptions and explanations of their subject matters are shaped by the origins and consequences of their research practices and by the interests, desires, and values promoted by such practices. How can the natural and social sciences be lead to take responsibility for their social locations and thus for their origins, values, and consequences? To ask this is to ask a social science question. Adequate *social* studies of the sciences turn out to be the necessary foundations upon which more comprehensive and less distorted descriptions and explanations of nature can be built. This conclusion is demanded by recognition that the culture "knows" a great deal that we individuals do not. The culture remains the "authoritative knower" of all those things about us for which we neglect or refuse personal and institutional responsibility. It "knows" the Eurocentrism and androcentrism that "natives" in the culture routinely express but cannot detect. If androcentric or Eurocentric beliefs and practices are part of the evidence for one hypothesis over another (inadvertently or not), then as part of scientific practice we must learn how to detect and eliminate them. Although the outcome of the natural sciences is shaped by how well this job is done, the methods of the natural sciences have been the wrong kind to do it. Consequently, it makes good sense to think of the natural sciences as a subfield of the critical social sciences. We will all have to think further about what this counter-intuitive proposal would mean in practice. Obviously, few fields of contemporary social science have methodologies, institutional structures, or agendas that are competent to identify the kinds of almost culturewide interests, values, and assumptions that end up functioning as evidence "behind the back" of the natural and social sciences, so to speak. Thus, it is one challenge to remedy this situation in the social sciences and another to conceptualize and then institutionalize a relationship between the natural and social sciences that will enable the former to get control of more of their evidence than they can now manage to do.

The Zairean philosopher V. Y. Mudimbe argues that just as European and American imperialists invented an Africa that would serve their purposes (they said they discovered it), so must Africans now invent a West that serves Africans' purposes. The imperialists claimed to discover in Africa a primitiveness, a prelogicality, an immorality that could serve as the opposite of the purportedly civilized West they were simultaneously inventing. But such an approach can be usefully developed by the other side as well, Mudimbe points out. For Africans today, he argues, a "critical reading of the Western experience is simultaneously a way of 'inventing' a foreign tradition in order to master its techniques and an ambiguous strategy for implementing alterity."[17]

The feminist discussions of science and epistemology are similarly engaged: we must "invent" the very Western sciences and institutions of knowledge in which we participate (and which pay some of our salaries) as bizarre beliefs and practices of the indigenous peoples who rule the modern West. We must master their techniques as we simultaneously continue to "discover" the ways in which they are "other" to ourselves and our agendas.

If we in the West can reinvent this part of the West, Western culture can learn things about itself and about the "others" against which it has built mighty conceptual and institutional fortresses. Of course, that will require different practices as well as different thoughts.

17. V. Y. Mudimbe, *The Invention of Africa: Gnosis, Philosophy, and the Order of Knowledge* (Bloomington: Indiana University Press, 1988), 171.

I

Science

2

Feminism Confronts the Sciences

Reform and Transformation

Only in the late 1970s did feminists begin to bring to bear on the theories and practices of science and technology the distinctive approaches that had been developing in the social sciences, the humanities, and, more generally, the women's movement. For many of us, it began to appear that in order to reform the sciences, as well as the philosophy, history, and social studies of science, there might have to be broad transformations of both science and society. Were reforms going to require revolutions? Were such revolutions going to be possible without further reforms?

Perhaps affirmative action in the natural sciences would require broader efforts and have more radical consequences than had at first been imagined. Perhaps more than the misuse or abuse of the sciences was at issue in the ways scientific technologies appeared in women's lives. Eliminating sexist bias in biology and the social sciences might require redefining objectivity, rationality, and scientific method. Eliminating the reliance on misogynistic metaphors in science could require eliminating gender itself if, as has been argued, gender is fundamentally an asymmetrical power relationship. In all these feminist approaches to science and technology, certain suspicions arose. Perhaps the fundamental problem was epistemological: within the conventional approaches to science studies, it was impossible to see how women's lives could be recognized as legitimate grounds from which true beliefs—or, at least, "less false" ones—could be generated. A "woman scientist" appeared to be a contradiction in terms; the reason for this was that "man scientist" named far too perfect a union.

Though similar epistemological issues emerged from all the different criticisms of the sciences and their technologies, there remain fundamental disagreements within the feminist writings, including the ways the epistemological issues are addressed. It is exactly because feminists approach issues about science and technology from the perspective of so many different projects that these criticisms raise some of the most interesting questions—and some of the most deeply challenging ones—about this central institution in contemporary Western social life.

Sometimes these criticisms are in tension with each other. Sometimes they are even internally contradictory. Their assumptions conflict with respect to what the goals of feminist criticisms of science and technology should be and how we should go about achieving any given goal. Such situations arise not because feminism is incapable of producing clear thought about science or because as women we cannot be expected to understand this "manly institution." These tensions, conflicts, and contradictions emerge because so many of the familiar assumptions about science that even feminists make are grounded only in the narrow, partial, and frequently unrealistic pictures of the scientific enterprise that prevail in conventional thought. When feminists borrow the most familiar approaches to science in order to bring about what one might think are relatively modest agendas, those agendas turn out to be unattainable. When they take more radical approaches in order to achieve those modest agendas, they frequently appear to be talking about things that conventional scientists and science observers do not even recognize as relevant or appropriate topics. As feminists have discovered in every field, when one tries to add women and gender to conventional subject matters and conceptual schemes, it quickly becomes obvious that the two have been defined against each other in such a way that they cannot be combined. The prevailing heroic dramas of the stalwart scientific iconoclast struggling against the dark forces of ignorance and superstition will not play right when women scientists (and/or scientists of color) step on the stage. Nor, we can note here, will the more mundane dramas produced by sociologists' and anthropologists' accounts of everyday laboratory life.

This chapter explores some of the issues through which conflicts, tensions, and contradictions emerge in the feminist approaches to the sciences and their technologies. To assist the reader in recognizing as wide a variety as possible of the central issues in the discussions of subsequent chapters, the following overview focuses on five areas: the analyses of women's situation in science, the sexist misuse and abuse of the sciences and their technologies, sexist and androcentric bias in the results of biological and social science research, the sexual meanings of nature and inquiry, and androcentric theories of scientific knowledge. I give the most weight to the first of these five focuses because women's situation in the social structure of the sciences is frequently the first issue of concern to scientists new to feminist analyses. I try to indicate a little more fully than in the other cases how an apparently simple issue—such as how to recruit more women into the sciences more effectively—quickly leads to deeper reflections on science and on society.[1]

One point first. Some of these issues concern both the social and natural sciences, but many will appear to be relevant only to the social sciences. Although that appearance is deceiving, this chapter will leave relatively untheorized (as it is in most writings about women and the sciences) the relationship between the metascientific and epistemological issues in the social sciences and those in the natural sciences. It is a theme of this book, however, that the relationship between the natural and social sciences is misleadingly theorized in general: that is, in the "prefeminist" metatheories of science and in various epistemologies. So there will be ample opportunity in later chapters to reflect on the consequences *for each other* of the feminist criticisms of the social and natural sciences.

Women in Science

First of all, among studies of women's lives within the social structure of science, there are at least four distinctive focuses. Each one has produced new understandings of women's situation, but taken alone,

---

1. I do not attempt to provide a literature review here. See Londa Schiebinger, "The History and Philosophy of Women in Science: A Review Essay," *Signs* 12:2 (1987); and Sandra Harding and Jean O'Barr, eds., *Sex and Scientific Inquiry* (Chicago: University of Chicago Press, 1987), for a representative sample of essays (including Schiebinger's illuminating review). The most recent and comprehensive bibliography is Alison Wylie, Kathleen Okruhlik, Sandra Morton, and Leslie Thielen-Wilson, "Feminist Critiques of Science: A Comprehensive Guide to the Literature," *Resources for Feminist Research, Canada* 90:2 (1990); see also the review of one part of this literature by the same authors, "Feminist Critiques of Science: The Epistemological and Methodological Literature," *Women's Studies International Forum* 12:3 (1989).

each one also limits our vision by its concentration on just one aspect of this issue. We need all four—and an analysis informed by the other concerns discussed below—to get a more comprehensive picture of the situation of women in the sciences.

Women Worthies

Historical studies and biographies of contemporary scientists bring to our attention the "women worthies" in science: the many women who have made important contributions but who are ignored or devalued in the androcentric mainstream literature. A new generation of historians is bringing to bear on the lives of these women the insights of several decades of feminist approaches to women's history. As they do so, we begin to get far richer analyses of the social resources and strategies (and the social processes outside their control) which have enabled women to achieve in these frequently hostile fields. Sciences have been hostile to women in some periods and welcoming in others, and these shifts have had as much to do with patterns in the development of individual sciences as they have with the changing fortunes of sexism and androcentrism.[2]

Class and race played roles in creating opportunities for these women, as they did for their brothers. When scientific collecting and experimentation were primarily gentlemen's activities, daughters as well as sons could gain a scientific education in the laboratory out behind the kitchen. It is striking how many early women scientists were related to male scientists; they learned science from or were supported in their work by fathers or husbands who were also scientists. Indeed, it is difficult in many eras to find women scientists who were not mentored by male relatives. Class and race opportunities are obviously related: poor women and women of color were as unlikely as *their* brothers to have relatives who were scientists; however, occasionally—very occasionally—their brothers did succeed in gaining access to science educa-

2. See Pnina G. Abir-Am and Dorinda Outram, eds., *Uneasy Careers and Intimate Lives: Women in Science, 1789–1979* (New Brunswick, N.J.: Rutgers University Press, 1987); Margaret Rossiter, *Women Scientists in America: Struggles and Strategies to 1940* (Baltimore, Md.: Johns Hopkins University Press, 1982); Londa Schiebinger, *The Mind Has No Sex: Women in the Origins of Modern Science* (Cambridge, Mass.: Harvard University Press, 1989).

tion, even if not to the kinds of careers to which such an education provided entrance for white men in the elite class. There is a real "Black Apollo" in recent American science; but "Black Athena" apparently can be an image only for the (important but now unjustly devalued) science and knowledge of premodern and non-Western cultures.[3]

The consciousness of the women who did find a place in science was often not feminist. Indeed, even a *woman's* consciousness could hardly be permitted if the fiction were to be maintained that a woman scientist must be a contradiction in terms. In order to succeed as scientists, these women usually had to force their lives as closely as possible into life cycles designed to accommodate the lives of men in patriarchal societies. Their possibilities for marriage and children were severely diminished in ways that never affected their brothers. Nevertheless, in both the nineteenth and twentieth centuries, many important scientists who were women have been active in projects everyone recognized as advanced by feminists. They have had to be, in order to open doors for themselves and their women students that were otherwise open only for male colleagues and their male students.

Moreover, we can ask whether "having a feminist consciousness" in the sense of overtly embracing feminism should really be regarded as a prerequisite for feminist activities: that is, for activities that have both the intention and the effect of specifically benefiting women. The term "feminism" is too radical for some people and too conservative for others. It is common today to find people struggling specifically to improve women's conditions but refusing to characterize their efforts as feminist, and to find different groups doing so for both progressive and conservative reasons. Furthermore, on the one hand, the success of the daily struggles for survival as a scientist in which these pioneers were forced to engage opened subsequent doors for women in the sciences. Whether or not they took part in overtly feminist activities, simply surviving against the odds was a remarkable achievement. "Never Meant to Survive" is the title given to one African American woman's account of her life in science in order to emphasize the diffi-

3. See Kenneth Manning, *Black Apollo of Science: The Life of Edward Everett Just* (New York: Oxford University Press, 1983); Martin Bernal, *Black Athena: The Afroasiatic Roots of Classical Civilization* (New Brunswick, N.J.: Rutgers University Press, 1987). Chapters 8, 9, and 11 investigate the benefits of really trying to "add race" to the feminist science criticisms and the interesting challenges that result.

culties for her of simply enduring in this institution.[4] On the other hand, it trivializes feminism to insist that every success achieved by women in a male-dominant institution is a cause for feminist pride. Feminism must have principles for distinguishing between feminism and self-interest (does "advocating feminist capitalism" have any meaning?), and between feminisms associated with progressive politics and those that may be progressively feminist in some respects but are or become infused with regressive tendencies in other respects—for example, with racism or class privilege. Perhaps we can at least admire survival, if not always for feminist reasons.

There is much more to be learned about science and women's situation in it by examining the lives of these women who were doubly pioneers—as scientists and as women (and triply so if they were women of color). With few exceptions they have been left out of the standard histories of science, engineering, mathematics, medicine, and the social sciences. There is much more to be done in restoring to historical awareness the record of these women's achievements and their struggles as women in the process. We do a disservice to our daughters and our sons—as well as to the historical record—in ignoring and undervaluing women's work. Yet far from being a phenomenon of the past, these exclusionary practices threaten to continue unless men as well as women are willing to speak up and fight against the androcentric biases that support such practices. After all, it was only twenty years ago that James Watson could devalue and ridicule in print—and with a macho hubris that signaled expected approval—the work of Rosalind Franklin in the discovery of the structure of DNA. It was Franklin's work as well as theirs that permitted Watson and Sir Francis Crick to win a Nobel Prize. Why was it not also awarded to her? Franklin was lucky enough to have a friend who was made so angry by Watson's account that she was willing to put in the time and energy to reconstruct the historical record and produce a book that corrected Watson's account.[5] Few of us—or our colleagues or daughters—will have friends with the knowledge available to Franklin's friend or the

4. Aimée Sands, "Never Meant to Survive: A Black Woman's Journey," *Radical Teacher* 30 (1986); the phrase is from a poem of Audre Lorde's, "A Litany for Survival," in *The Black Unicorn* (New York: Norton, 1978).

5. James Watson, *The Double Helix* (New York: New American Library, 1969); Anne Sayre, *Rosalind Franklin and DNA: A Vivid View of What It Is Like to Be a Gifted Woman in an Especially Male Profession* (New York: Norton, 1975).

willingness to commit a significant part of their lives to that kind of effort on our behalf.

The studies of "women worthies" if taken by themselves, however, can distort our understanding of the situation of women in science. Exactly because they were so unusual in their own day, the lives and experiences of these women are not typical. They do not tell us what we need to know in order to understand the experiences of the majority of women who try to make it into science or who may achieve less distinguished careers than these few. Great women's lives are in this respect no different from great men's lives, in science or elsewhere, which do not provide much illumination on the situation of ordinary people. Someone intending to teach or write about women's situation in science by generalizing from these women's lives would do a great disservice to the rest of the women in the sciences. This is especially true if the history of the great women is reconstructed uncritically from such conventional historical sources as, for instance, Watson's account of Franklin. But it is also true if the history is reconstructed primarily from the women's own accounts. They, like the rest of us, have been woefully unaware of the larger forces outside their immediate environments which have tended to shape their opportunities. Also, they have sometimes been unaware of the meanings that their achievements and their lives have had for other women. While they have often had to struggle mightily for admission to the sciences, their successes give them the status of "winners" in history's battles, who, historians point out, do not usually provide maximally objective accounts of the institutions in which they have succeeded. It requires critical social theory to understand the history of science, and such theory will have to be informed by the new race- and class- and gender-conscious social histories if it is to get beyond what the "natives" in science (women as well as men) can see.

The recovery of the histories of distinguished women scientists is important work, and much more of it is needed. But there is more we want to know about women in science than can be learned from this significant project alone.

Women's Contributions

Women's contributions to the history and practice of science are not limited to the achievements of a few extraordinary individuals. The new women's history and sociology have directed attention to the less public, less official, less visible, and less dramatic aspects of science in order to gain a better understanding of women's participation in these enterprises.[6] Salons organized and run by women enabled male scientists to find patrons to fund their laboratories and collecting trips in nineteenth-century France. In Europe and the United States, women's networks and more formally constituted women's clubs of botanists, biologists, and astronomers made important contributions to the collection of data. Some scientific fields, such as mathematics, have openly welcomed women in certain historical periods.[7] As elementary and secondary school teachers, women with advanced science training have provided a significant part of the preparation necessary for the subsequent pursuit of scientific careers by everyone in the scientific workforce, including the Nobel Prize winners. In their work as science and medical illustrators, editors, and popularizers, women have shaped perceptions that have made it possible to gather public support for scientific activity. Today, Nobel Prizes could not be won without the work of women as lab technicians and postdoctoral lab assistants, not to mention data analysts and computer programmers. Women have been active in the sciences in ways that are invisible if one focuses only on the public, official, visible, and dramatic figures and events that are favored in the conventional accounts.

In reading these new histories, one is reminded of the perception that comes to Virginia Woolf's Mrs Dalloway as she is preparing a dinner party to be attended by a bishop, several eminent statesmen, and various other figures significant in British social life early in this century. She thinks about how the British Empire would fall apart if women did not perform day in and day out the devalued and trivialized work that provides the social glue for the great achievements that men in the

6. See Marcia Millman and Rosabeth Moss Kanter, eds., *Another Voice: Feminist Perspectives on Social Life and Social Science* (New York: Anchor Books, 1975), for pointed early arguments about the importance of such concerns to an understanding of the organization of social life; and Margaret Andersen, *Thinking about Women: Sociological Perspectives on Sex and Gender*, 2d ed. (New York: Macmillan, 1988), for a comprehensive literature review and critical analysis of how these concerns have transformed sociological understandings of social life. The new women's histories and sociologies are complementary, but it is primarily the new historical perspectives that have been brought to bear in extended analyses of the sciences.

7. Abir-Am and Outram, *Uneasy Careers;* Rossiter, *Women Scientists in America;* Schiebinger, *The Mind Has No Sex.*

dominant social groups imagine they bring off all by themselves.[8] Moreover, it is primarily women—and especially of color and poor women—who provide far more than a fair share of the material conditions for these great achievements. They do the feeding, dressing, cleaning, nursing, and other kinds of care of elite men's bodies and the local places where those bodies exist—the relational and concrete labor that makes it possible for elite men to spend their days contemplating "all by themselves" the perfect motions of abstract, isolated bodies. Might our understanding of nature and social life be different if the people who discovered the laws of nature were the same ones who cleaned up after them?

This train of thought leads to skepticism about the value-neutrality—not to mention empirical adequacy—of the ways that "contributions to science" have been conceptualized in the mainstream philosophies, histories, and social studies of science. In the traditional studies, what counts as science and what counts as a contribution are determined by how elites in science and in society choose to define them. Feminists have argued that these definitions are self-serving (whether or not they are intended to be so) and that they obscure the important contributions of women in all classes and races to the production of whatever cultures count as their best kinds of knowledge.

One more issue pertinent to science studies is raised by Mrs Dalloway's perception. The conventionalists never ask what women's contributions to science have meant to women. What did it mean to the women involved and to others that it was women who gathered so much of the data upon which theories depended in biology, botany, and astronomy? What has it meant for women that they devote their lives as daughters, wives, or mothers to the production and maintenance of men who direct the institution of science? What does it mean for women in the sciences and the rest of us today to recover the history of these contributions? I suggest that feminists today are ambivalent about the meanings for us of these histories, as were perhaps at least some of our foremothers, for similar reasons. On the one hand, we delight in finding that women made important contributions to the advance of an institution that our society values so highly. On the other hand, we are not sure that the institution should be quite so highly

8. Virginia Woolf, *Mrs Dalloway* (1925; New York: Harcourt Brace Jovanovich, 1964).

valued, especially in light of its often destructive consequences, or that so much of the value assigned to women's activities should be for their contributions to the welfare of men who often show little regard for the welfare of women. Restricting our understanding to the meanings that contributions to science have for those who write the heroic dramas about science deprives us of this more complex perspective. Did Mrs Dalloway doubt that it was a good thing to advance the interests and concerns of elites in that one segment of the modern Western empire? Many people certainly do.

Structural Obstacles

From the beginning, feminist observers of science have critically examined the structural obstacles to women's achievement in science. Historical studies of the formal barriers to women's equity reveal the vigorous campaigns that were usually necessary to allow women the access men had to scientific education, degrees, lab appointments, publication in journals, membership in scientific societies, jobs in universities or industry or the government, and scientific prizes.

Historical studies of the emergence, flourishing, and eventual decline of the formal barriers to women in the sciences have been complemented by sociological and psychological studies of informal barriers. Motivation and psychoanalytic studies have shown why boys and men more often want to enter and remain in science, engineering, and math than do girls and women.[9] For one thing, the very same personality traits that young males must take on to become masculine in the modern West are just those that are particularly valued for careers in science and related fields. Facility in abstract thought, physical interaction with the environment, and a conception of nature as separate and in need of control—which parents and the society encourage in male children in order to make them more manly—are just what prepares young people to like and excel at math, science, and engineering. Correlatively, in order to make female children feminine and womanly, parents encourage a tendency toward concrete and relational thought and a preference for personal, caring service to other people. These traits prepare girls and women to prefer teaching, mothering, and other

9. See, e.g., Violet B. Haas and Carolyn C. Perrucci, eds., *Women in Scientific and Engineering Professions* (Ann Arbor: University of Michigan Press, 1984).

28

service and caring activities to those that are essential for careers in mathematics, science, and engineering.[10]

Moreover, sociological studies of the mechanisms of informal discrimination against women have shown how much harder it is for women than for men to reinvest the capital of credentials that they do accumulate in a way that ensures equal advancement of their careers. The devaluation of any work known to have been done by women, the exclusion of women from men's informational networks, the obstacles put in the path of women's attempts to obtain safe and reliable mentors (and, later, to be perceived as such mentors themselves)—these and other informal discriminatory tactics give us increased appreciation for those women who have managed to persist.[11]

One must emphasize that structural obstacles should be the focus here—not the purported biological or personality traits on which the sexist attempts to explain women's lack of equity in science have concentrated. Nor should the focus be primarily on the sexism of individual men or even precisely of men as a group. James Watson's treatment of Rosalind Franklin was certainly sexist, and he should be held responsible for that behavior. But of more interest for analyses of the structure of the sciences and attempts to change the future for women in the sciences is his correct perception that neither men nor women would criticize his behavior (at least, not until feminism came along). His bragging tone reveals the existence not just of one little sexist but, more important, of the entire institution of gender relations, including male supremacy, that he could depend upon to support his beliefs and behaviors.

10. Studies of Western thought that are indebted to psychoanalytic theory and illuminating for thinking about science and epistemology include Isaac Balbus, *Marxism and Domination* (Princeton, N.J.: Princeton University Press, 1982), esp. chaps. 9–10, discussing the correlation between cultural differences in child-rearing practices and cultural attitudes toward nature; Nancy Chodorow, *The Reproduction of Mothering* (Berkeley: University of California Press, 1978); Jane Flax, "Political Philosophy and the Patriarchal Unconscious: A Psychoanalytic Perspective on Epistemology and Metaphysics," in *Discovering Reality: Feminist Perspectives on Epistemology, Metaphysics, Methodology, and Philosophy of Science*, ed. Sandra Harding and Merrill Hintikka (Dordrecht: Reidel, 1983); Nancy Hirschman, "Freedom, Recognition, and Obligation: A Feminist Approach to Political Theory," *American Political Science Review* 83:4 (1989), 1227–44; Evelyn Fox Keller, *Reflections on Gender and Science* (New Haven, Conn.: Yale University Press, 1984); and Evelyn Fox Keller and Christine R. Grontkowski, "The Mind's Eye," in Harding and Hintikka, *Discovering Reality*.

11. See, e.g., Rossiter, *Women Scientists in America;* Haas and Perrucci, *Women in Scientific and Engineering Professions*.

Thus, "personal bias" is not the issue here—irritating as it is to experience the bias of individuals. Preferring strawberry ice cream, refusing to wear a tie, or hating television are quite different matters from advancing (intentionally or not) the political agendas of the dominant groups in a society. Personal biases are idiosyncratic and of little scientific or institutional importance; that is why they are called "personal." Sexism and androcentrism are culturewide social, political, and economic characteristics, and it obscures their origins, institutionalization, and consequences to refer to them as personal biases.

An exclusive focus only on the structural obstacles to women's equity would give the impression that science as an institution as well as the individual men in science have been totally resistant to women's scientific activities, and that women have rarely succeeded in this institution. Of course, neither is the case. This kind of preoccupation with "victimology" must be balanced with the studies of women's resistance to marginalization and their achievements.

Science Education

Finally, attention to improved science, math, and engineering education for girls and women is one important concern of more general educational and curricular studies. This research is used to shape programs that try to create environments and learning processes that will encourage young women, and not just young men, to enter and remain in such fields.[12]

Another improvement in science education is less often discussed than it should be, both inside and outside feminist circles: what should we teach young—and not so young—men and women about how science functions? If "science education for girls" means the same kinds of educational opportunities and supportive environments available to their brothers, the implication is that boys' science education is just fine. But the argument of this book is that such an assumption is damaging to science, men, women, and society. An implication of the

---

12. See Sue Rosser, *Teaching Science and Health from a Feminist Perspective* (New York: Pergamon Press, 1986); Sue Rosser, *Feminism in the Science and Health Care Professions: Overcoming Resistance* (New York: Pergamon Press, 1988); Joan Rothschild, *Teaching Technology from a Feminist Perspective* (New York: Pergamon Press, 1988).

rest of the feminist criticisms of science and technology (discussed below and in later chapters) is that the conventional scientific education available to boys is woefully flawed in that it is structured by unrealistic and politically damaging images of and goals for scientific activity. Such a view offers opportunities for radical thought about what science education should look like for everyone.

Some of these issues are on the agendas of educators concerned to bridge C. P. Snow's "two cultures." They are also the concern of those who wish to provide the kind of education that will, for example, keep physicians and science researchers alert to the need to stay out of lawsuits. Even such an august body as the National Academy of Sciences is insisting on the understanding that scientific work is always and necessarily shaped by social values, an understanding that would have been unimaginable to most scientists and philosophers of science only a decade ago.[13] The purpose of the academy's document is to "uphold the integrity of science" against fraud, misallocation of credit, plagiarism, and other such damages to scientific "property," but the image of science-in-society used to do so is in some respects quite radical. What should we want students to know about the scientific enterprise, its history, practices, and goals? Would any of them want to go through the arduous training necessary to become a scientist if they were told the truth, the whole truth, and nothing but the truth about this institution and its present-day practices?

All four of these ways of examining the situation of women in science raise challenges to the practices of the scientific enterprise down through history and to certain aspects of the liberal, progressive image of science favored by many scientists and government officials and in much scholarly and popular rhetoric—though the tide is beginning to shift toward greater acknowledgment of the ways the social order structures sciences's institutionalized practices. Nevertheless, the focuses on equity for women have been regarded by many as the least radical of the criticisms of science because they appear not to challenge either the logic of the inquiry process or the logic of scientific explanation. We have already begun to see how the exploration of issues that at first appear extremely conservative or merely reformist quickly moves us toward far more radical issues. Issues of access for women to the

13. National Academy of Sciences, *On Being a Scientist* (Washington, D.C.: National Academy Press, 1989).

practice of science turn out to have substantive scientific and epistemological consequences as well as moral and political ones.

Moreover, the sheer documentation of discrimination against women in the social structure of science draws attention to the mystifying effects of science's claims to universalism. In this social institution alone, we are told, the social identity of its members is irrelevant to achievement. It is irrelevant to the "goodness" of the results of research whether researchers are Japanese or British, white or black, Catholic or Jewish. Science can afford the luxury of such tolerance, it is said, because scientific method is powerful enough to eliminate from the results of research any social values that might have crept into the scientific process through the social values of individual scientists. But the history of vigorous individual and institutional attempts to exclude women as well as virtually every other category of "others"—Jews, blacks, other people of color—from positions of status and authority in science shows that claims to universalism are in fact intended to apply only within the dominant gender, classes, races, and cultures. Feminism highlights the hypocrisy and irrationality of these universalistic claims in the face of overt and tacit discriminatory practices. And it is the tendency in recent feminism regarded by many as least radical—liberal feminism—that has this radical consequence. In trying to bring women as a "sex class" into domains that have been reserved for men, liberal feminism reveals that it is only men whom liberal theory could have had in mind when it proclaimed the rights of individuals (indeed, only men of the preferred class and race).[14] This is a harsh judgment—no doubt many will think it too harsh by far—and certainly science is by no means the only institution to exhibit these unattractive characteristics. But we will let the judgment stand in order to stimulate more critical thought about the gap between, on the one hand, the traditionally glorious and progressive images science calls upon when seeking material and political support for its activities and, on the other hand, the reality that emerges when we take seriously the facts about the practices of science raised by feminist critiques, among others.

Even though these critiques will turn out to have radical consequences for the logic of inquiry and explanation, it has to be admitted

14. Zillah Eisenstein, *The Radical Future of Liberal Feminism* (New York: Longman, 1981); Hirschman, "Freedom, Recognition, and Obligation."

that many women in science who make the equity claims do not offer a challenge to the existing social structure or politics of the natural and social sciences. This is a problem. Should women want to become just like men in science, as many of these studies appear to assume? To which men in science should women want to be equal? Presumably, the answer is not underpaid and exploited lab technicians, or men in racial minorities who have also suffered exclusion and devaluation.

Moreover, it is convenient to overlook the deep ties between science and warmaking. The new histories of science, such as Margaret Jacob's, bring these connections into sight, revealing that

> Western science at its foundations, as promoted by its most brilliant as well as its most ordinary exponents, never questioned the usefulness of scientific knowledge for warmaking. I know of no text from the early modern period which suggests that the scientist should withhold his knowledge from any government, at any time, but especially in the process of preparing for warmaking. Indeed most texts that recommend science also propose its usefulness in improving the state's capacity to wage war more effectively, to destroy more efficiently.[15]

Late twentieth-century physics, too, has been shaped by military control, as studies of U.S. science policy since World War II show.[16] Why should feminists want women or anyone else to be military researchers—or to engage in apparently valuable kinds of scientific or medical research whose results will be distributed only for profit? Neither the internal social structure of science nor science's involvement in exploitative politics is deeply challenged by the ways in which the equity issues are often formulated. Further, little attention has been given to the special achievements of and obstacles faced by women of color.[17]

There are good reasons to want more women in physics and other natural sciences, as Chapter 3 shows. But it is hard to discover them in

15. Margaret Jacob, *The Cultural Meanings of the Scientific Revolution* (New York: Knopf, 1988), 251.

16. Paul Forman, "Beyond Quantum Electronics: National Security as the Basis for Physical Research in the U.S., 1940–1960," *Historical Studies in Physical and Biological Sciences* 18 (1987).

17. Exceptions can be found in the bibliography by Anne Fausto-Sterling and Lydia English, "Women and Minorities in Science: Course Materials Guide," 1982 (available from Fausto-Sterling, Division of Biology and Medicine, Brown University, Providence, RI 02912).

much of the conventional literature intended to attract girls into science education and keep women in scientific jobs. Much of this recruitment literature has the numbing, depressing, and even alarming feel of U.S. military recruitment literature for the Vietnam war.

The critical examinations of the situations of women in science, then, have their problems—problems that originate in the distorting conventional understandings of the institution of science and its social functions. Nevertheless, this focus of feminist concern has the potential to generate radical changes in science. Even before discussing that potential, we can already see that women's achievements in this extremely masculine enterprise can gain status and power for women in worlds previously forbidden them. The meanings of these accomplishments for women and for men throughout society are already radical. Moreover, these achievements gain for women valuable skills and information about the natural and social worlds. They increase women's scientific literacy, and in today's world, that is as important as it is to teach women to read and write.

Sexist Misuse and Abuse of Science and Technology

Studies of the uses and abuses of biology, the social sciences, and their technologies show how they have been used in the service of sexism, racism, homophobia, and class exploitation. One important feminist focus in such research is on reproductive technologies and policies. Within this literature are studies of historical and contemporary aspects of the medicalization of birth, the introduction of reproductive technologies that often do not benefit women, sterilization and abortion policies, pronatalism, unsafe contraceptives, unnecessary gynecological surgery, and many other issues. But whatever the topic, common themes emerge. It is clear that the dominant culture has been willing to take far greater risks with women's reproductive systems than it would ever countenance for those of men—or, rather, for those of men in the dominant classes and races. This is just one way of saying that women's lack of power in the social order extends to their having far less control than do men in the dominant groups over what happens to their own bodies. Moreover, feminist analyses of reproductive issues again and again raise deeply disturbing questions for everyone about fundamental assumptions in Western thought, such as the inevitable benefits of greater knowledge and of greater choices.[18]

Not only technologies and applied sciences but also scientific theories have been used to move control of women's lives to those who exercise power in the dominant class, race, and culture.[19] Many egregiously sexist and androcentric misuses and abuses have been documented for workplace and domestic technologies.[20] And gender relations more generally—not just those that take the form of male supremacy—are implicated in the applications of science that result in ecological destruction and support militarism.[21]

Fruitful conceptual tensions and conflicts emerge in these writings, as well as between their ideas and conventional ways of thinking about the misuse and abuse of the sciences and technologies. These analyses challenge the idea that any and all scientific, medical, and technological research should be supported with public funds on the grounds that all such research increases knowledge and resources *for humanity*. In the first place, it is clear that the sciences and their technologies today, as in the past, provide such benefits disproportionately to members of the dominant races, classes, and gender. Moreover, the resources are frequently used not just for the benefit of the few but also for the direct oppression and exploitation of the many. As critic of

---

18. See, e.g., Rita Arditti, Renate Duelli-Klein, and Shelly Minden, eds., *Test-Tube Women: What Future for Motherhood?* (Boston: Pandora Press, 1984); Patricia Spallone and Deborah Steinberg, eds., *Made to Order* (New York: Pergamon Press, 1987).

19. Many critics make this argument. For one good recent analysis, see Ruth Hubbard, *The Politics of Women's Biology* (New Brunswick, N.J.: Rutgers University Press, 1990).

20. See Cynthia Cockburn, *Machinery of Dominance: Women, Men, and Technical Know-How* (Boston: Northeastern University Press, 1988); Barbara Ehrenreich and Deirdre English, *For Her Own Good: 150 Years of Experts' Advice to Women* (New York: Doubleday, 1979); Ruth Schwartz Cowan, *More Work for Mother: The Ironies of Household Technology from the Open Hearth to the Microwave* (New York: Basic Books, 1983); and critical studies of the situation of women in Third World development projects: e.g., Susan C. Bourque and Kay B. Warren, "Technology, Gender, and Development," *Daedalus* 116:4 (1987); Maria Mies, *Patriarchy and Accumulation on a World Scale: Women in the International Division of Labour* (Atlantic Highlands, N.J.: Zed Books, 1986); Maria Mies, Veronika Bennholdt-Thomsen, and Claudia von Werlhof, *Women: The Last Colony* (Atlantic Highlands, N.J.: Zed Books, 1988).

21. H. Patricia Hynes, *The Recurring Silent Spring* (New York: Pergamon Press, 1989); Sara Ruddick, *Maternal Thinking: Toward a Politics of Peace* (Boston: Beacon Press, 1989).

science William Leiss noted, science's claim to seek to dominate nature in order to control "man's fate" has actually hidden its real function and, often, intention: now and in the past, whether scientists intended it or not, science has provided resources for some people's domination of others.[22] Is it possible that *more* scientific, medical, and technological research in societies stratified by race, class, and gender actually *increases* social stratification? Perhaps not, but the question certainly needs to be critically examined.

In the second place, there is the criticism that science *is* a social problem because the society that shapes it is a social problem. In Sal Restivo's words, "Modern science is the product of an alienated human spirit and a factor in rather than a solution to our individual and social problems."[23] The sciences and their technologies are not at all "autonomous" from the social order in important respects: they emerged as part of and remain today an integral component in modern social relations. From this perspective, maintaining funding for the sciences reproduces individual and social problems rather than contributing to solving them.

These critiques can bring to the surface of the feminist science discussions the class, race, and imperialist projects of the West in which the sciences and their technologies have been and remain deeply implicated. Thus, they are radical in this way, too. They force feminist observers of science to notice that "women" are always women of a certain race, class, and culture. We are white as well as black; economically overprivileged as well as underprivileged; Western as well as Third World—and race supremacy, economic overprivilege, and Eurocentrism are problems that the sciences have helped to advance. That is, when race is the topic of discussion, white is at least as appropriate a focus as black. We replicate the oppression characteristic of androcentric discourses if we fail to observe that scientific and technological benefits accumulate for Westerners, whites, and the economically overadvantaged as the correlative disadvantages accumulate for Third World peoples, racial "minorities," and the poor.

Nevertheless, this focus, like that on women in the social structure of the sciences, can appear to leave the pure core of science untouched by feminist challenges. After all, the science enthusiast can say, these crit-

22. William Leiss, *The Domination of Nature* (Boston: Beacon Press, 1972).
23. Sal Restivo, "Modern Science as a Social Problem," *Social Problems* 35:3 (1988), 206.

icisms are applicable only to technologies and to the applications of science; pure science, the theoretical science that wins Nobel Prizes, is not implicated in the misuses and abuses of science and technology in the political realm. As this argument goes, we can have an easy conscience when we teach our students pure theoretical science. We and they are not responsible for what happens to the value-neutral information that is the result of pure scientific inquiry once it leaves the hands of scientists. After it is released into society, how it is used for good or bad purposes becomes others' responsibility.

The appropriate response to that kind of defense reveals yet another radical aspect of this feminist critique. The very language of "misuse and abuse," borrowed from conventional science studies, in fact gives away too much to the defenders of pure science. The sciences are part and parcel, woof and warp, of the social orders from which they emerge and which support them. "Science versus society" is a false and distorting image. First, social desires are frequently defined as technological needs, and as such they generate and legitimate scientific research for both individuals and institutions. Social desires have been defined as technological needs for better firearms, better navigation for sea voyages between Africa and the Americas, machines for large-scale and more cost-efficient industrial and agricultural production, and more effective means of controlling population growth. The "problematics" of science frequently arise from the translation of social agendas into technological ones.

Second, the technologies used to produce scientific information are not value-neutral. For example, the development of the telescope moved authority about the patterns of the heavens from the church to the secular world and supported the emerging importance of the authority of individual observation. Contemporary scientific technologies—computers, laboratory tests, nuclear accelerators—shift values in the sciences in other ways.[24]

Third, and most obvious, the sciences generate information that is used to produce technologies and applications that are not morally and politically neutral. The institution of science hides this fact in a variety of ways, most notably by constantly splitting off into what it conceptualizes as separate fields those sciences in which a focus on

24. See, e.g., Stanley Reiser, *Medicine and the Reign of Technology* (Cambridge: Cambridge University Press, 1978); Sharon Traweek, *Beamtimes and Life Times: The World of High Energy Physicists* (Cambridge, Mass.: Harvard University Press, 1988).

application has become obvious. (Is the moment of obviousness also the moment of redefinition as "separate"?) Thus electrical, mechanical, and civil engineering are "not really physics"; medicine, health policy, nursing, and the invention of reproductive technologies are "not really biology"; "applied mathematics" and "chemical engineering" have been sloughed off to separate disciplines, disciplinary organizations, journals, and university departments.

These efforts to absolve scientists and the groups supporting scientific research of responsibility for the consequences of science are not always successful, as many of the Manhattan Project physicists, for example, came to understand. There is no such thing as "pure science," and there probably never was. After all, Galileo wrote in Italian, not the Latin favored for scholarly interchanges in his day; and he worked in the arsenal in Venice, not in a monastery. Both individual scientists and the institutions of science must be assigned responsibility for their locations in history, whether or not they actively chose those locations. They must be expected to learn the predictable consequences of their beliefs and behaviors. Scientific institutions should be required to exhibit the concern for the causes and consequences of their own beliefs and behaviors that they insist on for everyone else's.

This issue brings us full circle to an increased appreciation of the question of why the sciences we have deserve public support. Why *should* society, in the face of competing social needs, provide massive resources for an enterprise that claims in itself to have no social consequences? There is a vast irrationality in this kind of argument for the purity of science. One has a sense that to even question it is to risk falling to a different reality in which the conceptual framework for the dominant science discourses appears as bizarre to us as did the townspeople's beliefs to the child in the fable about the emperor and his new clothes.

There is one problematic aspect of some feminist responses to the "misuse and abuse" critiques. We need to think about the possible positive effects for women of technologies that initially appear to offer only disasters and consider how to bring about these good effects. For example, it is certainly understandable that many feminists have been staunchly opposed to new reproductive technologies which, in the present misogynous cultures, threaten to decrease the already deteriorated control women have over their bodies and their reproductive choices. Moreover, the potential of these technologies for increasing class and race exploitation is clear. Nevertheless, we need to develop more discriminating responses to them because the chances of stopping them entirely are low, they do offer benefits to some women, and many of these women do not now have access to them: arguments against the hospital-centered delivery of health care are not especially appealing to women whose poverty and lack of insurance denies them and their children access to hospitals and doctors. Women are facing only the newest form of familiar questions about the most appropriate and most effective means—not always identical—for gaining control over technologies that shape their lives.

Sexist and Androcentric Bias

Researchers have raised a variety of criticisms against the sexism and androcentrism that have shaped the results of research in biology and the social sciences. We can now see that females as well as males have made important contributions to the evolution of our species. Contrary to Darwinian and other interpretations of evolutionary theory, females too have evolved.[25] We can see that Woman the Gatherer played at least as significant a role as did Man the Hunter in ushering in the dawn of our species. Woman's moral development now appears different but just as necessary for the conduct of social relations—not immature, deviant, and a "problem" for the maintenance of social order.[26] Women's contributions to social life begin to appear once the paradigms of the social that seem natural to male sociologists and their male informants are enriched with understandings of women's less visible, dramatic, and public but nonetheless important activities.[27] In history, economics, and linguistics, too, much of what has passed for distinctively human activity and belief can now be seen to be only masculine activity and belief—and distinctively modern, Western, and upper-class at that.

What are some of the general conceptual issues that these criticisms raise about science and its dominant images?

25. Ruth Hubbard, "Have Only Men Evolved?" in Harding and Hintikka, *Discovering Reality.*
26. Carol Gilligan, *In a Different Voice: Psychological Theory and Women's Development* (Cambridge, Mass.: Harvard University Press, 1982).
27. Andersen, *Thinking about Women;* Millman and Kanter, *Another Voice.*

Androcentric biases can enter the research process at every stage, as critics have shown.[28] They enter in the concepts and hypotheses selected, in the design of research, and in the collection and interpretation of data. The most radical implication of understanding how the structure of the institution of science structures the content of the science produced there, however, may be the recognition that whoever gets to define what counts as a scientific problem also gets a powerful role in shaping the picture of the world that results from scientific research.

The biases that enter science in the form of problematics—the definition of certain phenomena as problematic—are impossible to eliminate through the kinds of controls that biologists and social scientists think of as research methods. This finding challenges the old and widespread claim that one must make a sharp distinction between the "context of discovery" and the "context of justification." It makes no difference, philosophers and scientists say, whether one's scientific problems arrive as a result of sun worship (as Kepler's did) from talking to astrologers and alchemists (as many of the early modern scientists did), or from talking to the best scientists of the day (where these are *not* the astrologers and alchemists); social values undoubtedly arrive in science from all of these discovery contexts. But once hypotheses are in the context of justification where they can be subjected to rigorous testing, the argument continues, all social values can be detected and eliminated, leaving only pure facts—information—as the results of research.

In contrast, critics have argued that the difficulty is not just that the picture of nature and social relations generated by the sciences is shaped in large part by what individual scientists happen to think of as a scientific problem. It is that what gets to count as a problem is linked with the purposes for which research is done—or at least for which it is funded. If the dominant gender perceives women's sexuality as problematic, we will get one view of human sexuality; if men's is perceived as the problem, we will get another. If explaining gender inequality without implicating powerful men or the rightness of male supremacy is implicitly defined as the scientific problem, one picture of human biology, history, and social life emerges; if the problem is to explain

28. See, e.g., the analysis by Helen Longino and Ruth Doell, "Body, Bias, and Behavior: A Comparative Analysis of Reasoning in Two Areas of Biological Science, *Signs* 9:2 (1983).

gender inequality and let the chips fall where they may, a different picture will emerge. If the problem is defined as overpopulation in the Third World, one scientific agenda will emerge; if it is the refusal of the West to permit Third World cultures to retain the resources they need to support their populations, a different agenda will seem appropriate. The sciences' methods of research will do little to eliminate social biases intrinsic to these conflicting definitions as long as only one of the two problems is defined as a scientific one. And after centuries of primarily dominant groups defining what gets to count as scientific problems, do we need an equally long history of subordinated groups making the majority of such definitions? If not, how is our distorted picture of the world to be corrected?

One way to see this issue is to notice that though scientific methods are selected, we are told, exactly in order to eliminate all social values from inquiry, they are actually operationalized to eliminate only those values that differ within whatever gets to count as the community of scientists. If values and interests that can produce the most critical perspectives on science are silenced through discriminatory social practices, the standard, narrowly conceived conception of scientific method will have not an iota of a chance of maximizing either value-neutrality or objectivity.

Such a conclusion has the effect of turning equity issues into scientific and epistemological issues, not only moral and political ones. Members of an oppressed group are more likely than are their oppressors to be suspicious of false claims made about the oppressed group. The conservative defense of equity claimed that more women in science would not change the nature of the science produced. It may well be true that more women in a science and a society still highly [woman's voice] committed to male supremacy in other respects would have little effect on the content and purposes of science. But the discussion here suggests that our picture of the world would indeed be different if, in the context of a woman's movement and of feminist theory, women's voices were powerful in shaping the general direction of scientific research in biology and the social sciences. Of course men, too, can learn to listen to women's voices and to generate scientific agendas that are—intentionally or not—less self-serving. Some already have. So the issue is not exclusively one of "which bodies" are represented in the scientific community. However, there is no reason at all to think that antidemocratic communities can produce maximally objective knowledge. Democracy and science are deeply linked in ways I will explore further.

One more issue to be pursued later should be raised at this point. If feminists are critical of scientific method, is there an alternative feminist method of research that has a better chance of producing unbiased results? The moment one tries to answer this apparently perfectly reasonable question, such confusions emerge from its formulation that no simple, unproblematic answer is possible. In a certain sense, there is a distinctive feminist "method" of research—a distinctive way of beginning, of finding the most fruitful questions, of grounding knowledge claims. But in another sense, the very desire for a method of inquiry—a technique, formula, algorithm, or intellectual mechanism—that can guarantee objective accounts of women's lives and the social order should be regarded with suspicion. Feminism is a politics too, and commitment to that politics cannot be separated from feminist standards for what should count as reasonably produced claims to knowledge.[29]

The Sexual Meanings of Nature and Inquiry

Science produces information, but it also produces meanings. Indeed, as even some conventional philosophers of science realize, the results of scientific research *are* information only if they have meaning for us; an undecipherable string of numbers or nonsense syllables is not yet information. Moreover, science produces meanings of its own activities which are intended to create resources for it. It leads us to think of its kind of method as a moral good, as the place where the inherently positive value of science is to be found—hence the term

29. This method issue has been a source of confusion among feminists. See, e.g., Sandra Harding, ed. *Feminism and Methodology: Social Science Issues* (Bloomington: Indiana University Press, 1987); Sandra Harding, "The Method Question," *Hypatia* 2:3 (1987) (a different version of the introduction to the edited collection). See also "Commentary by Naomi Scheman," "Comment by Dorothy Smith" (on the *Hypatia* essay above), and Sandra Harding, "Response" (to Scheman and Smith), *American Philosophical Association Newsletter on Feminism and Philosophy* 88:3 (1989). Dorothy Smith objected to my criticism of feminists and others for trying to conceptualize political changes in the sciences in terms of changes in method. I should have noted in the original essay that Smith has been radically transforming the term "method" in an important way (her work is discussed esp. in Chapter 5).

"positivism." Science produces meanings of itself as a "calling," as one of the most important supports of the rational life, as a heroic struggle, as the paradigm of a distinctively human activity, as the apogee of civilization, and so on and so on. It produces meanings of its methods of research as stripping the veils from nature, or torturing nature to reveal her secrets, or, more attractively, as attempting to defeat "might makes right" in the domain of empirical knowledge. And science delivers meanings for the nature it describes and explains—as something requiring domination or hiding its secrets, as a worthy opponent, even as the bride of the scientist.[30]

Hence, another focus of the feminist examination of science uses techniques of literary criticism, historical interpretation, and psychoanalysis to "read science as a text." The text is the whole of science: its formal statements, intellectual traditions, research practices, social formations, the scientific and popular beliefs about it, and so on. Francis Bacon appealed to rape metaphors to persuade his audience that experimental method is a good thing: "For you have but to hound nature in her wanderings and you will be able when you like to lead and drive her afterwards to the same place again. Neither ought a man to make scruple of entering and penetrating into those holes and corners when the inquisition of truth is his whole object."[31] Paul Feyerabend, a contemporary philosopher of science, has recommended his own analysis over competing ones by saying that "such a development . . . changes science from a stern and demanding mistress into an attractive and yielding courtesan who tries to anticipate every wish of her lover. Of course it is up to us to choose either a dragon or a pussy cat for our company. I think I do not have to explain my own preferences."[32] In his Nobel Prize acceptance speech, physicist Richard Feynman called the idea that inspired the work for which he won his prize an "old lady, who has very little that's attractive left in her, and the young today will

30. See, e.g., Morris Berman, *The Reenchantment of the World* (Ithaca: Cornell University Press, 1981); Brian Easlea, *Witch Hunting, Magic, and the New Philosophy* (Brighton, Eng.: Harvester Press, 1980); Brian Easlea, *Fathering the Unthinkable* (London: Pluto Press, 1983); Keller, *Reflections;* Leiss, *Domination of Nature;* Carolyn Merchant, *The Death of Nature: Women, Ecology, and the Scientific Revolution* (New York: Harper & Row, 1980).

31. Quoted in Merchant, *Death of Nature,* 168.

32. Paul Feyerabend, "Consolations for the Specialist," in *Criticism and the Growth of Knowledge,* ed. Imre Lakatos and Alan Musgrave (New York: Cambridge University Press, 1970), 229.

not have their hearts pound when they look at her anymore. But we can say the best we can for any old woman, that she has become a very good mother and has given birth to some very good children. And I thank the Swedish Academy of Sciences for complimenting one of them."[33]

When we realize that the mechanistic metaphors that organized early modern science themselves carried sexual meanings, it is clear that these meanings are central to the ways scientists conceptualize both the methods of inquiry and models of nature. Restrained but clear echoes still appear even in a text that is clearly trying to keep science "buttoned up": "The laws of nature are not apparent in our everyday surroundings, waiting to be plucked like fruit from a tree. They are hidden and unyielding, and the difficulties of grasping them add greatly to the satisfaction of success."[34] Such metaphors and gender meanings of scientific methods, theories and objects of knowledge rescientize sex stereotypes as they simultaneously generate distortions of nature and inquiry.

These criticisms raise a number of important issues.[35] For one thing, it is important to see that the focus should not be on whether individuals in the history of science were sexist. Most of them were; in this they were like most men (and many women) of their day. Instead, the point is that the sexual meanings of nature and inquiry are used to express the anxieties of whole societies—or, at least, of the groups whose interests science was intended to advance. Cultural meanings, not individual ones, should be the issue here. Appeals to familiar sexual politics are used to allay anxieties about perceived threats to the social order.[36]

Moreover, the interactionist theories of metaphor that have been developed in the last few decades make clear that these metaphors are not merely heuristic devices or literary embellishments that can be replaced by value-neutral referential terms. They are a substantive part of science in that they show scientists how to extend the domains of

---

33. Richard Feynman, *The Feynman Lectures in Physics* (Reading, Mass: Addison-Wesley, 1964). See Sandra Harding, *The Science Question in Feminism* (Ithaca: Cornell University Press, 1986), chap. 5, for further discussion of these metaphors.

34. National Academy of Sciences, *On Being a Scientist,* 6.

35. See Harding, *The Science Question in Feminism,* 233–39.

36. See Merchant, *Death of Nature;* and Susan Bordo, *The Flight to Objectivity* (Albany: State University of New York Press, 1987).

their theories, what regularities of nature they should expect to find, what questions about nature to ask.[37]

These sexist meanings are politically and morally obnoxious. But they also distort our understandings of nature in two ways. For one thing, if scientists tend to select (intentionally or not) certain kinds of methods of inquiry *because,* among other reasons, they are associated with distinctively masculine stereotypes—interventionist methods, for example—masculine stereotypes have become part of the *evidence* for the results of this research. Since we should be able to weigh all the evidence for a scientist's claims, this preference for certain methods on the grounds that they carry masculine meanings or avoid feminine ones should be presented as part of the evidence. (Imagine such a research report in a scientific journal!)[38]

One consequence of the prevalence of this sort of preference is that scientists become less able to understand those aspects of nature that are not detectable through such methods and models. For instance, if Barbara McClintock's noninterventionist observation of patterns of growth in corn is associated with distinctively nonmasculine styles of interaction, it will be less used and appreciated by people who over-value masculinity and devalue femininity.[39] In the second place, partiality for macho approaches to nature can also distort explanations. If hierarchical models of causation and control are associated with desirable masculine personality traits, the less hierarchical aspects of nature will be harder to detect, because they are not given reality, made visible, by the preferred hierarchical model. For example, in studies of slime mold aggregation, the imposition of such hierarchical causal notions as that of a pacemaker has made it difficult to see the interactive aspects of these processes.[40] Emily Martin has pointed out that modern Western medicine regards the female body as a kind of factory that derives its fundamental value from the quantity and quality of its products—that is, babies; once this factory is no longer able to manufacture these products, it is conceptualized as obsolete, useless. This

37. Mary Hesse, *Models and Metaphors in Science* (Notre Dame, Ind.: University of Notre Dame Press, 1966).

38. For an illuminating discussion of how social and cultural values shape evidence, see Helen Longino, *Science as Social Knowledge: Values and Objectivity in Scientific Inquiry* (Princeton, N.J.: Princeton University Press, 1990).

39. See, e.g., Keller, *Reflections,* chap. 9.

40. Ibid., chap. 8.

metaphoric view builds in both capitalist and androcentric values. It reduces women to their reproductive functions, and it makes difficult any understanding of female bodies as capable any other kinds of contributions to the social order, let alone any other uses or values *for* women themselves.[41]

Should we want these metaphors eliminated from science? Where possible, this would clearly appear to be desirable. But because these metaphors have become attractive in scientific work as a result of their social uses outside science, eliminating them from the language of science can be only part of the solution to the problem of how to degender the natural sciences. Doing so cannot in itself end the practice of drawing upon social meanings in the direction of research. Moreover, it raises the issue of how one should evaluate obviously sexist metaphors that have, nevertheless, contributed to the growth of scientific knowledge. How should feminism analyze, for instance, the fact that mechanistic metaphors drew on misogynous politics as a resource for the development of science?

In a certain sense, these are the wrong questions. Obviously, the sexist language of science is continuous with sexist thought in society in general. All thought and language both shapes and is shaped by the social order, its projects, and attempts to resolve conflicts within it. So solutions to this problem cannot be independent of more general struggles to end the subjection of women, racial "minorities," and the poor and to transform sciences into knowledge-seeking institutions *of, by,* and *for* these groups. We need to ask how to create the kinds of societies in which the dominant institutions of knowledge production are no longer so complicitous in benefiting the few to the detriment of the many. This critique is especially important however, because it shows that physics, chemistry, and abstract thought in every realm (including philosophy) can be deeply sexist or androcentric even when no humans at all appear in their domain of inquiry. Evidently, abstract thought is not quite as abstract as most have assumed. Perhaps even excessive preferences for the abstract themselves undercut the point of abstraction: these preferences, like all others, can be historically located.

41. Emily Martin, *The Women in the Body* (Boston: Beacon Press, 1987).

Androcentric Epistemologies

Finally, the preceding challenges to the natural and social sciences raise epistemological issues. One way to put the problem here is to note that in the humanities, biology, and the social sciences, it turned out to be impossible to "add women" without challenging the foundations of those disciplines. Similarly, the feminist science critics argue that when we try to "add women" as knowers within traditional theories of knowledge, we quickly discover how partial and distorted are those theories.

The issue is not whether individual women have gained scientific trainings and credentials and have made important contributions to the growth of knowledge; thousands and thousands of them have done so. The issue for the feminist epistemological critiques is a different one: "woman the knower" (like "woman scientist") appears to be a contradiction in terms. By "woman the knower" I mean women as agents of knowledge, as actors on the stage of history, as humans whose lives provide a grounding for knowledge claims that are different from and in some respects preferable to knowledge claims grounded in the lives of men in the dominant groups.

We can begin to sense the contradictions when we note that conventionally, what it means to be scientific is to be dispassionate, disinterested, impartial, concerned with abstract principles and rules; but what it means to be a woman is to be emotional, interested in and partial to the welfare of family and friends, concerned with concrete practices and contextual relations. Feminists have argued that these features of womanliness are not the consequences of biology—let alone of inferior biology. Rather, they arise from a variety of social conditions that are more characteristic of women's lives than of the lives of men in the dominant groups. One argument is that men in the dominant groups assign to women (and other marginalized peoples) certain kinds of human activity that they do not want to do themselves. They assign to women the care of all bodies, including men's, and the local places where bodies exist (houses, offices, and so on), the care of young children, and "emotional work"—the processing of men's and everyone else's feelings. (Some men will do emotional labor only if they can charge psychiatrists' fees for it.) In short, most dominant-group men refuse to be responsible for daily life, their own or other

47

people's.[42] Must women renounce what they can know about nature and social relations from the perspective of their daily lives in order to produce what the culture is able to recognize as knowledge? How can such socially situated knowledge be justified when the dominant epistemology assumes that since real knowledge is transcendental, the very idea of situated knowledge is a contradiction in terms?[43]

Distinctively feminist analyses of theories of scientific knowledge have begun to crystallize around three different and partially conflicting approaches. First, feminist empiricism attempts to bring the feminist criticisms of scientific claims into the existing theories of scientific knowledge by arguing that sexist and androcentric results of research are simply the consequence of "bad science." From this perspective, feminists are helping science to better follow its widely recognized procedures and to achieve its existing goals. Second, the feminist standpoint theorists argue that the problem is more extensive. The dominant conceptual schemes of the natural and social sciences fit the experience that Western men of the elite classes and races have of themselves and the world around them. Political struggle and feminist theory, they say, must be incorporated into the sciences if we are to be able to see beneath the partial and false images of the world that the sciences generate. By starting research from women's lives, we can arrive at empirically and theoretically more adequate descriptions and explanations—at less partial and distorting ones. Finally, a third approach argues that even these two feminist theories of knowledge are not radical enough. Both still adhere too closely to damaging Enlightenment beliefs about the possibility of producing one true story about a reality that is out there and ready to be reflected in the mirror of our minds. These postmodernist tendencies in feminist thought point to the far too intimate connections between science and power in the past. They ask whether feminist epistemology is to continue the policing of thought which characterizes the conventional epistemology-centered philosophies and sciences.

42. Dorothy Smith, *The Everyday World as Problematic: A Feminist Sociology* (Boston: Northeastern University Press, 1987); Nancy Hartsock, "The Feminist Standpoint: Developing the Ground for a Specifically Feminist Historical Materialism," in Harding and Hintikka, *Discovering Reality;* Bettina Aptheker, *Tapestries of Life: Women's Work, Women's Consciousness, and the Meaning of Daily Life* (Amherst: University of Massachusetts Press, 1989).

43. See Donna Haraway, "Situated Knowledges: *The Science Question in Feminism* and the Privilege of Partial Perspective," *Feminist Studies* 14:3 (1988).

The chapters ahead argue for a postmodernist standpoint approach that is nevertheless committed to rethinking and reusing some important notions from conventional metatheories of science. That is, I follow the logic of the standpoint approaches into postmodernist terrains while trying, en route, to refashion familiar but incompetent conceptual utensils into useful ones for everyday work in the sciences, philosophy, and democratic struggles of the present moment. Others are engaged in similar projects, though they may not describe them in these terms.

Thus epistemological issues weave through all of the chapters that follow. What are the differences, if any, between an epistemology and a justificatory strategy? Are the feminist epistemologies really philosophies of knowledge, or are they only sociologies of knowledge? Are they too indebted to the patriarchal theories from which they borrow—empiricism, the standpoint of the proletariat, the dominant tendencies in postmodernism? Or, to the contrary, do they really undermine these very resources on which they nevertheless draw? Exactly what are the grounds in women's lives claimed by standpoint epistemology? Are they essentialist? Can they speak to the differences in women's experiences, activities, lives? What is the relationship between these feminist approaches and other contemporary tendencies toward greater democracy? Are these feminist theories of knowledge still excessively loyal to Western conceptions and evaluations of science? Are they relativist? Are they regressively foundational? Do they abandon objectivity? Or do they overvalue science? Does feminist politics damage or strengthen them? Can men be feminist knowers? These and other questions are pursued below.

In *The Science Question in Feminism* I characterized the journey from issues about women in science to criticisms of theories of scientific knowledge as a shift from the "woman question in science" to the "science question in feminism." In its concern with equity issues, the woman question begins by asking, "What do women want from science?" It conceptualizes women as a special interest group—like, for instance, farmers or oil producers in their relationship to the government—who ask that their special needs and interests be fairly recognized in the institution of science. By the time we get to the epistemology discussions, both the issues and how women are conceptualized have radically shifted. The science question asks, "How can we use for emancipatory ends those sciences that are apparently so intimately involved in Western, bourgeois, and masculine projects?" And women appear not as a special interest group pleading for a hearing for their own interests alone but as thinkers expressing concerns about science and society that are echoed in the other "countercultures" of science—in antiracist and Third World movements, in anticapitalism movements, and in the ecology and peace movements. This should not be surprising: there are feminists in all these other movements, and many participants in these other movements are feminists.

These shifts reflect the radical potential of feminism. It is both reformist and revolutionary; conventional political dichotomies do not capture its most important tendencies. Some social observers have suggested that it is the most recent of the great revolutions of modernity to sweep across the cultures of the world. Struggles against the exploitation of women take their place in a history shaped also by struggles against the exploitation of the working classes and of peoples of color. But a chronological placing of feminism's tendencies may be more confusing than helpful: because all forms of exploitation are irretrievably interlocked, so too must be the struggles against them. Time will tell how important and successful feminism becomes as a social movement; this is not the first moment in history to see a "rising of the women," as the nineteenth-century feminists called their struggles. Meanwhile, the feminist challenges to science raise important new questions for all of us about the histories we have had and the futures we should want for the relationship between knowledge and the social order.

3

How the Women's Movement Benefits Science

Two Views

The methodology and epistemology of modern science assume that people are interchangeable as knowers. "Anyone can see through my telescope," said Galileo. He, like the later members of the "New Science Movement" that flourished in England in the seventeenth century, intended to take out of the monopoly of the aristocrats, the humanists, and the priesthood the right to state with good evidence what are the regularities and underlying causal tendencies of the natural world.[1] "Science for the people," he proclaimed, a phrase that was taken up again in the 1960s for its implications of a science that could advance participatory democracy.[2]

But from the perspective of groups that society excludes and marginalizes, this now conventional claim that all knowers should be interchangeable can appear to have certain antidemocratic consequences. If all knowers are interchangeable, then affirmative action in the sciences can be "only" a moral and political agenda. It can have no possible positive consequences for the content or logic of the natural sciences; the scientific work of men and women, blacks and whites, Nazis and Ku Klux Klanners will be equally supervised and disciplined by scientific method. If all knowers are in principle interchangeable, then

1. Wolfgang Van den Daele, "The Social Construction of Science," in *The Social Production of Scientific Knowledge,* ed. Everett Mendelsohn, Peter Weingart, and Richard Whitley (Dordrecht: Reidel, 1977).
2. See the journal *Science for the People.* Galileo's intentions were more complex than these allusions suggest. See, e.g., Margaret Jacob, *The Cultural Meaning of the Scientific Revolution* (New York: Knopf, 1988).

white, Western, economically privileged, heterosexual men can produce knowledge at least as good as anyone else can. They speak for nature itself and not, as their critics suspect, out of any particular historical assumptions about nature, how humans should interact with nature, or human social relations for which empirical knowledge could be a resource. Thus affirmative action, and the civil rights and women's movements that have struggled so heroically to establish and maintain affirmative action agendas in universities and workplaces, appear to have no relevance to the picture of the world produced by the sciences or to our theories of knowledge.

Even worse, from the conventional perspective it can appear more likely that affirmative action policies will endanger the sciences. If one assumes (contrary to the reports of most observers of science as a social institution) that the members of society who have the greatest ability to "do science" are already being identified by and effectively recruited into the sciences, where the industrious exercise of their talents is alone responsible for advancing them as far as they can go, then the call to give special consideration to the recruitment of women and minorities can appear to endanger this purported meritocracy. It introduces irrelevant and damaging considerations into the sciences' recruitment policies. Moreover, while men in the dominant race and class are presumed to enter the scientific enterprise with only or primarily universal interests—they intend to "do good science," to discover nature's secrets, and so on—women and minorities are assumed to act far more often out of self-interest. And the demands by the latter for admission into science indicate that their activities are likely to be excessively shaped by politics rather than by the pursuit of pure information, which, it is thought, shapes the activities of men in the dominant groups. People who enter the sciences not as individual inquiring minds but as members of social groups with political agendas (whether or not the individual members of groups such as "women" and "minorities" are themselves committed to these agendas) are more likely to bring into their work distorting social values—or so the paradoxical logic of interchangeable knowers goes. Their work must be even more closely supervised and disciplined than the work of those who enter as disembodied individual minds: that is, as committed only to "pure science." Think of Lysenkoism and of Nazi science, they say.

Thus, it is even possible for people to weigh the costs of affirmative action against the costs of greater vigilance against distorting social values and to decide that, all in all, affirmative action agendas should be resisted. The women's movement, which agitates for affirmative action, appears to these people to be a negative influence against which the sciences must protect themselves if they are to preserve their own agendas. While I have tried here to make this line of reasoning as plausible as possible, I must point out at the start that it is hard to resist attributing a certain element of paranoia to people who equate with Stalinism and Nazism the entrance of a few more African Americans and women into an institution so overwhelmingly directed by white males, both inside science itself and in the state and industrial councils where so many research agendas are generated. One is reminded of the fear of the "feminization of science" decried by nineteenth-century scientists when women first began to enter their ranks.[3]

This chapter looks at the positive results that affirmative action agendas and other consequences of the women's and civil rights movements can have for the sciences and for knowledge-seeking more generally. Perhaps it is finally time to give up once and for all a political, moral, and epistemological position that was developed in its original form to move Europe out of feudalism.

Feminist Critiques: Solidarities and Differences

One focus in recent writings by feminists working in and around the natural sciences has been on the resources the women's movement offers to science. Three such resources, it has been argued, make unique contributions to the growth of knowledge in the natural sciences: women scientists, feminist politics, and feminist theories about science. Feminists who think about these resources are in both solidarity and conflict with each other. They agree that the women's movement does make important contributions to the growth of knowledge, but they disagree about just what those contributions are.

The solidarity is created by the necessity of working together against conventionalists, who regard as deeply threatening to some of their most cherished assumptions the very idea that a social movement could make contributions to the growth of knowledge. Even a few scientists

---

3. See Margaret Rossiter, *Women Scientists in America: Struggles and Strategies to 1940* (Baltimore, Md.: Johns Hopkins University Press, 1982), e.g., 313.

and observers who think of themselves as radical critics of science in other respects are conventionalists in this respect. They think that their own social action on behalf of workers, Third World peoples, the environment, antimilitarism, or animal liberation makes important contributions to the growth of knowledge but that the feminism the women's movement has produced does not. On a daily basis, then, all feminists in and around the sciences have to struggle together against individuals and institutions that say or act as if women are not likely to become really good physicists, chemists, mathematicians, or engineers, and imagining that science can learn anything from the women's movement is guaranteed to generate only regressive, distorting, and repressive political ideology and a decline in the quality of scientific activity—despite clear cases in which the feminist perspective *has* contributed to the growth of knowledge.

However, feminists who must struggle together against these conventionalists disagree among themselves about other important issues. One such disagreement separates those who believe that the task of feminist analysis is to object to "bad science" from those who think that "science-as-usual"—the whole scientific enterprise, its purposes, practices, and functions—should be the target of feminist criticism. Indeed, feminists from one group are often surprised to discover just how different their agendas are from those of people in the other group who also think themselves feminists. The critics of bad science sometimes see the critics of science-as-usual as undermining the former's attempts to end sexist hiring practices and sexist and androcentric biases in the sciences, and as raising issues about race, class, and imperialism that seem to them to have only marginal relevance to the advancement of (white, Western) women in (white, Western) science. On the other side, the critics of science-as-usual sometimes see the critics of bad science as a distracting and particularly difficult part of the problem with science-as-usual. They see the feminist critics of bad science as complicitous with our culture's failure to question deeply enough the ethics, goals, and functions of science. A few feminists want to explore the relationship between these agendas, and others simply draw on whichever agenda seems appropriate at the moment, without worrying about conflict between them. But many feminists concerned with science find compelling only one or the other of these two general approaches.

I find this situation troubling for a number of reasons. On the one hand, if science-as-usual is the problem, then it appears that feminism should not encourage more women to become scientists and thus part of this problem. But it is distressing that an apparent consequence of the success of feminist criticisms of this field would be to alienate women from entering it—especially when women were for so long vigorously excluded from it by patriarchal culture and when many women have waged such heroic campaigns to enter and remain there. It is not a purpose of the feminist criticisms of literature that women should stop writing, or of feminist criticisms of the social sciences that women should abandon attempts to understand the social world. How bizarre if it is an outcome of the purportedly most radical feminist science criticisms that women should give up trying to understand the natural world. Can we try to create more feminist sciences without any women scientists? Moreover, we live in a scientific culture; to be scientifically illiterate is simply to be illiterate—a condition of far too many women and men already. It is regressive tendencies in science-as-usual that foster such illiteracy, as feminist critics have pointed out. But if women do not become scientists, would not that fact further discourage girls and women from becoming scientifically literate and from gaining the kinds of control of our own lives that such literacy makes possible? Should feminism join science-as-usual in fostering scientific illiteracy among women? What could be progressive about that?

On the other hand, it is equally troubling that many of the critics of bad science are resistant to examining the fact that the social structure and purposes of contemporary science are created far away from scientists' daily experience; they are created as part of the bourgeois, racist, imperialist—as well as androcentric—policies of the ruling groups in society. It is an important principle of that social order to maintain the separation between the laboratories where scientists experience science and the councils where science policy is made. Should feminism want women to have equality with the men of their respective races and classes without challenging race and class exploitation? Should feminism want women, too, to do research that will predictably be used to mount military attacks on small Third World societies or to increase profit for the few? What is progressive about organizing heroic campaigns to "add women and gender" to the social structure and subject matters of the sciences without questioning the legitimacy of science's social hierarchy and politically regressive agendas?

Fortunately, there are now more and more discussions between critics of bad science and critics of science-as-usual. It is crucial that this dialogue be informed by the voices of the majority of the world's women who are involved not at all in criticizing the sciences but simply in surviving. The desire to eliminate, control, and economically exploit them has set far too many of the agendas of science-as-usual. How can they benefit from natural sciences that are supposed to be for the improvement of "humanity"?

One topic of these discussions should be to envision now how the emerging feminist natural sciences should develop.[4] Feminist theories that contain alternative conceptions of nature, social relations, and ways to obtain less partial and distorted knowledge of the empirical world are already shaping scientific research projects in the social sciences and some areas of biology. If feminists and women around the world could choose the structure and projects for physics, chemistry, engineering, and biology, what would we choose? To begin to answer this question wisely requires dialogue between the critics of bad science, the critics of science-as-usual, and the rest of the world's women. Such discussion is already a priority in women's and development studies and the women's health movement, but they could profitably include a more comprehensive view from women's lives. (I realize that the question I pose seems exactly what the traditionalists imagine in their nightmares that feminist science critics are preoccupied with: "What would a feminist physics be like?" Perhaps we should take more of our cues from the traditionalists; their nightmares could lead us to some of our most progressive projects!)

Identifying the outlines and origins of difference between the agendas of the two main feminist approaches to science can clarify their dialogue. I try to do this by sketching briefly what I consider to be their central features,[5] then looking at their contrasting positions with respect to how resources for science are provided by women in science, feminist politics, and feminist theories about science. (The third major feminist theory of knowledge, referred to as feminist postmodernism or feminist poststructuralism, is studiously uninterested in advancing feminist sciences; see Chapter 7 and Part III.)

---

4. I now think it important to say that there *are* already feminist sciences (as I argue in Chapter 12), a position I did not defend in the essay of which this chapter is a revision.

5. Part II takes up in more detail the epistemological assumptions that structure these two approaches.

The Critique of "Bad Science"

As noted in Chapter 2, feminist researchers in biology and the social sciences have shown in convincing detail the sexist and androcentric results of research that does not carefully enough follow well-understood principles of method and theory. Basing generalizations about humans only on data about men violates obvious rules of method and theory. Failing to question why women's responses to moral dilemmas do not comfortably fit categories designed to receive men's responses should long since have been regarded as unreasonable.[6] Research designs that legitimate having only men interview only men about either men's or women's beliefs and behaviors are bound to distort reality. Assumptions that women's reproductive systems normally function in immature or pathological ways lack grounding in principles of biology, not to mention in common sense.

If critics ended their arguments with such attacks on bad science (most have not), they would be attempting to preserve principles of good research and the logic of explanation (the logic of justification, in philosophers' terms) as these are widely understood today in the natural and social sciences. This understanding was developed at the moment when the natural sciences first became positivist.[7] But science is older than its positivist era. Only in the late seventeenth century was it first said that the positive benefit of science could be restricted to its method, thus making it unnecessary for scientists and the institution of science to be overtly concerned with the social, political, and economic origins, consequences, or constituting values of science. (The term "positivism" is an even later invention; it named an idea that was already well understood.) Those who see the problem only as bad science support the goal of value-neutral objectivity and impartiality for all scientific inquiry; they assume that there is an Archimedean vantage point from which the relations of the natural and social world can appear in their proper perspective. Opinions are divided about whether one should discuss the remnants of positivism under that name. Some natural scientists, many social scientists, but almost no philosophers of science will happily describe their own philosophy of science as positivist. Other observers are quite sure that no one at all is

6. Carol Gilligan, *In a Different Voice: Psychological Theory and Women's Development* (Cambridge, Mass.: Harvard University Press, 1982).
7. See Van den Daele, "Social Construction of Science."

really a positivist any more, so to criticize positivism (or "excessive empiricism," as some of us have called it) is only to criticize straw figures.[8]

In the conventional theory, the subject of knowledge—the scientist, the knower—is always an individual; the knower cannot be a group such as a social class or a gender. And this individual is abstract; "it" cannot have any particular historical social identity. The authors of the results of scientific research are supposed to be socially anonymous; it should not make any difference to the "goodness" of the results of research whether the researcher is Chinese or British, African American or European American, a woman or a man. Scientific method is supposed to be powerful enough to eliminate any social biases that might find their way from the social situation of the scientist into hypotheses, concepts, research designs, evidence-gathering, or the interpretation of the results of research.

Though its defenders rarely recognize it as such, this theory of knowledge is in fact part of a world view that includes as other central elements liberal political and moral theory. The Archimedean observer of good science is the impartial administrator of liberal political theory and the disinterested moral philosopher—the "good man"—of liberal ethics.[9] I have called the feminist form of this theory, as applied to science and its procedures for producing knowledge, "feminist empiricism."

The Critique of Science-as-Usual

The second major theory about science borrows the structure of a Marxist epistemology.[10] Knowledge is grounded in particular, histor-

---

8. See Roy Bhaskar, *Reclaiming Reality* (New York: Verso, 1989), chap. 4, "Philosophies as Ideologies of Science: A Contribution to the Critique of Positivism," for a delineation and critique of the tenets of positivism still widely held by science enthusiasts.

9. Alison Jaggar, *Feminist Politics and Human Nature* (Totowa, N.J.: Rowman & Allenheld, 1983). See also Jaggar, "Love and Knowledge: Emotion in Feminist Epistemology," in *Gender/Body/Knowledge*, ed. Alison Jaggar and Susan Bordo (New Brunswick, N.J.: Rutgers University Press, 1989).

10. See Nancy Hartsock, "The Feminist Standpoint: Developing the Ground for a Specifically Feminist Historical Materialism," in *Discovering Reality: Feminist Perspectives on Epistemology, Metaphysics, Methodology, and Philosophy of Science*, ed. Sandra Harding and Merrill Hintikka (Dordrecht: Reidel, 1983); Hilary Rose, "Hand, Brain,

ical social situations. In societies where power is organized hierarchically—for example, by class or race or gender—there is no possibility of an Archimedean perspective, one that is disinterested, impartial, value-free, or detached from the particular, historical social relations in which everyone participates. Instead, each person can achieve only a partial view of reality from the perspective of his or her own position in the social hierarchy. And such a view is not only partial but also distorted by the way the relations of dominance are organized. Further, the view from the perspective of the powerful is far more partial and distorted than that available from the perspective of the dominated; this is so for a variety of reasons. To name just one, the powerful have far more interests in obscuring the unjust conditions that produce their unearned privileges and authority than do the dominated groups in hiding the conditions that produce their situation. The unearned privileges for the few are paid for by the unjust misery of the many. As Hegel formulated the point that structures this theory, masters see the activity of slaves as an expression either of the slave's "nature" or of the master's will. From the perspective of the slaves, the situation looks very different. The two perspectives are not scientifically or epistemologically equal.[11]

Thus, in this theory the subject of belief and of knowledge is never simply an individual, let alone an abstract one capable of transcending its own historical location. It is always an individual in a particular social situation, and so in this sense it is also the social group that shares that situation. I always see the world through my culture's eyes; I think within its assumptions. In other words, our society can see only through our eyes and think its characteristic thoughts with our minds. There is no impartial, disinterested, value-neutral, Archimedean perspective. Nevertheless, it is possible to make reasonable judgments that some beliefs are better supported by empirical evidence than others. No one can tell the one, eternally true, perfect story about the way the world is; but we can tell some stories about ourselves, nature, and

and Heart: A Feminist Epistemology for the Natural Sciences," *Signs* 9:1 (1983); Dorothy Smith, *The Everyday World as Problematic: A Feminist Sociology* (Boston: Northeastern University Press, 1987); and my discussion of Smith's work in *The Science Question in Feminism* (Ithaca: Cornell University Press, 1986) and in Part II of this book.

11. Additional reasons for this scientific and epistemic asymmetry are discussed in Chapter 5.

social life which can be shown with good evidence to be far less partial and distorted—less false—than the dominant ones.

Although this theory of science, like the critique of "bad science," has emerged from research in biology and social science, its challenge to the practice of grounding scientific problematics, hypotheses, concepts, research designs, evidence, interpretations, and purposes of research, only in the lives of men of the dominant groups has the consequence of challenging the natural sciences as well. As I asked earlier, what would natural sciences look like if they were driven by purposes of research and problematics originating in women's lives and identified by feminism? One could hold that there are two branches of science with different "logics" of inquiry and explanation: in the social sciences and parts of biology, the social origins of research can play a positive role in the growth of knowledge; in the natural sciences, they have no role to play. Indeed, this is the way many philosophers and scientists have resolved the problem.[12] But such a position is questionable for a number of reasons. For one thing, almost all natural science research these days is driven by technology. Scientists may not be motivated by visions of new technologies of control or for profit, but funders of scientific research are. Feminist critics who challenge not bad science but science-as-usual are challenging the fit of science—past and present—with the gender, race, and class projects of its surrounding culture.

The two approaches to science can be thought of as ways to shape a "science of science," since one goal for both is to describe and explain the regularities and underlying causal tendencies of the practices of science. I say "one goal" because these approaches have been formulated not primarily as an intellectual exercise but in an urgent effort to figure out how to do research that does not have the bad effects for women that are characteristic consequences of conventional modes of scientific thought and practice. I have sketched only the broad outlines of these two feminist approaches to science, but they are sufficient to reveal why one finds within feminist discussions alternative analyses of the resources available to science in women scientists, feminist politics, and feminist theories about science. The two approaches shape different interpretations of the value of the women's movement to science.

12. An overview of these issues can be found in Brian Fay and Donald Moon, "What Would an Adequate Philosophy of Social Science Look Like?" *Philosophy of Social Science* 7 (1977).

Criticisms of Bad Science

How Women Scientists Benefit Science

For the critics of bad science, justice demands that women be given the same opportunities as their brothers for education, degrees, lab appointments, publication, teaching positions, membership in professional societies, awards, and the other benefits that participation in science can provide. The perceived radicalness of this apparently modest principle of equal opportunity becomes evident the minute one looks at the heroic struggles that have been necessary to eliminate the formal barriers against women's equality in science, mathematics, and engineering. Similarly vigorous struggles are still going on; they are unlikely to become unnecessary as long as informal barriers to equal opportunity remain successful. The issue is not that there are few women in science, for there are vast numbers of women with science degrees working in the scientific enterprise. The issue, instead, is why there are so few women directing the agendas of science.[13]

The grounds for the demands for equal opportunity have always been that women can do good science just as well as men and that they should be given the same opportunities to demonstrate their abilities. Women are an overlooked segment of the "manpower" pool from which science draws its workers. Thus, it could be said that women scientists' contribution is simply that they enlarge the pool of talented persons who can become good scientists. From this perspective, there would appear to be no political or conceptual space to argue that women scientists have a *special* contribution to make, as women, to the growth of scientific knowledge.

The critiques of bad science clarify why women in the natural sciences are resistant to the possibility that they should somehow be doing science *as women* rather than as the impartial, disinterested, value-neutral observers and thinkers that they were trained to be and that this feminist empiricism legitimates. The women's movement creates the possibility for more women to be scientists by attacking the formal and informal barriers that make it difficult for women to gain

---

13. Rossiter, *Women Scientists in America;* Violet B. Haas and Carolyn C. Perrucci, eds., *Women in Scientific and Engineering Professions* (Ann Arbor: University of Michigan Press, 1984); Londa Schiebinger, *The Mind Has No Sex: Women in the Origins of Modern Science* (Cambridge, Mass.: Harvard University Press, 1989).

the opportunities in science that are available to their brothers. But there is no claim here that women scientists provide any special resources *as women* to the growth of scientific knowledge. Consequently, the claim other feminists have made about women's abilities to make special contributions to the growth of knowledge *should* be troubling to women who, in order to achieve their precarious positions, have had to insist that they were "just one of the boys," that in the labs they certainly were not functioning as women. Whatever maternity leaves, child care, or other accommodations to women's reproductive and family roles they might think it appropriate to ask for, the critiques of bad science tell them that the way they do science and the content of their work is not, and should not be, affected by the fact that they are women. In this respect, the critics of bad science deviate not at all from the principle of abstract individualism that grounds conventional theories of science.

How Feminist Politics Benefits Science

The position becomes a little more complex when we ask whether, in addition to generating more scientific "manpower," the women's movement provides other benefits to science. One could argue that presumably the quantitative change discussed above has qualitative results: adding to the number of good scientists in the workforce should have a positive effect on the growth of scientific knowledge. If this is not true, should we infer that cutting the number of male scientists by half would have little effect on the growth of knowledge? But I want to pose a different question: does the women's movement contribute to the growth of knowledge by transforming the content or logic of science?

A reply given by two empirical sociologists, Marcia Millman and Rosabeth Moss Kanter, states with particular clarity the position of the critics of bad science:

> Everyone knows the story about the Emperor and his fine clothes: although the townspeople persuaded themselves that the Emperor was elegantly costumed, a child, possessing an unspoiled vision, showed the citizenry that the Emperor was really naked. The story instructs us about one of our basic sociological premises: that reality is subjective, or rather, subject to social definition. The story also reminds us that collective delusions can be undone by introducing fresh perspectives.

Movements of social liberation are like the story in this respect: they make it possible for people to see the world in an enlarged perspective because they remove the covers and blinders that obscure knowledge and observation. In the last decade no social movement has had a more startling or consequential impact on the way people see and act in the world than the women's movement. Like the onlookers in the Emperor's parade, we can see and plainly speak about things that have always been there, but that formerly were unacknowledged. Indeed, today it is impossible to escape noticing features of social life that were invisible only ten years ago.[14]

Echoing themes from the sciences' conventional self-image are the claims that the child has "unspoiled vision" in comparison to the townspeople; that unspoiled vision brings "fresh perspectives"; that the townspeople "persuaded themselves" to believe a delusion; that the women's movement "removes blinders" from our eyes, enabling us to "see things" that have always been there but were not visible to us earlier.

Although the conventional theory of how good science is done is still presented in technicolor here, it appears to be slipping out of focus in spite of the authors' intent. Evidently, scientific claims do have historical authors, since sometimes their reality is and sometimes it is not revealed by movements for social liberation, and claims revealed by such movements are scientifically better than those they would replace. Before the women's movement, our eyes were covered; the women's movement opens them for us. The assumptions of abstract individualism—that I can see reality all by myself, that scientific method alone is powerful enough to remove the blinders that obscure knowledge and observation—begin themselves to appear a foolish delusion. But this passage still suggests nothing special about women *as women* for science to use as a resource. Men and women can both lose their blinders as they learn from the unspoiled vision and fresh perspectives the women's movement provides.

Moreover, it does not appear necessary for scientists to go out of their way to engage in the politics of the women's movement in order to increase the kinds of benefits that the movement brings to science.

14. Marcia Millman and Rosabeth Moss Kanter, Editors' Introduction to *Another Voice: Feminist Perspectives on Social Life and Social Science* (New York: Anchor Books, 1975).

Someone has to do such politics for these benefits to science to accrue, but that someone doesn't have to be a scientist. Feminism appears available as a discourse and as an identity to people of good intentions if they open their minds to what the women's movement reveals to us all; if they get rid of their superstitions, ignorance, and prejudices; and, of course, if they have a humane attitude toward the disadvantaged and participate when asked in attempts to gain justice for them. Feminist science here is no more than the science that such people—men and women—already do. And in the case of physics, chemistry, and parts of biology, it hardly seems worthwhile to attach the label "feminist" to the science done in the presence of a women's movement. Maybe physicists will have to speak up on equity issues and watch their language a little more carefully to avoid offensive sexist metaphors. But nothing fundamental to how description and explanation of the natural world are produced will be done differently from the ways in which sciences are practiced when no women's movement is around. I am not arguing that Millman and Kanter would, in fact, make these claims but only that the unconsciously held but widespread theory of science that their passage echoes implies these claims.

Feminist Theory about Science as a Scientific Resource

Finally, what do the critics of bad science have to say about the role of social theories of science as a resource *for* science? Very little. If pressed, these thinkers tend to be reluctant to own that they have a theory of science at all. The excessively empiricist theory of science that they draw on presents scientific research as following a formula or algorithm. "For my way of discovering sciences goes far to level men's wits, and leaves but little to individual excellence because it performs everything by surest rules and demonstrations," says Bacon of his method.[15] Doing good science requires little reflection on how hypotheses come to be proposed or considered appropriate for testing, let alone on how social forces might make *positive* contributions to the growth of knowledge. This position goes back to Newton's refusal to admit that he even made hypotheses: "I frame no hypotheses; . . . hypotheses, whether metaphysical or physical, whether of occult

15. Quoted in Van den Daele, "Social Construction of Science," 34.

qualities or mechanical, have no place in experimental philosophy."[16] This position holds that the products of the mind (in contrast to the products of "nature") constitute obstacles for science; hypotheses and theories, like all products of the mind, should be regarded with suspicion.

Supporting these views is the assumption, widespread in the sciences and in philosophy, that while false beliefs often require social explanations, true beliefs are the consequence only of natural processes. Hence the "spoiled vision" of the townspeople deserves an explanation that refers to the social causes of the "spoiling," but the "unspoiled vision" of the child can be explained entirely with reference to natural causes: he is a child. Recently, this assumption has been the target of criticism from the "strong programme" in the sociology of knowledge, whose proponents call for causally symmetrical accounts of both true and false, legitimate and illegitimate beliefs. Otherwise, these sociologists point out, the sociology of knowledge is really only the sociology of error or of people legitimated as knowers.[17] The critics of bad science appear ambivalent about whether the explanations of the results of good science research should or should not refer to social causes, but (my point here) they do not think it important to doing good science to have a distinctive social theory of how to do good science. For example, they do not give their implicitly held theory of science a name.[18]

The Millman and Kanter passage quoted above is the closest to a statement of the benefits feminist social theory can bring to science that I have found in this literature. Presumably, its authors would say not just that movements of social liberation enlarge the vision available to science, whether or not scientists realize it, but also that the *understanding* of the positive effect of such social values on the growth of knowledge—an understanding advanced by the very fact that they write this passage—should be useful to science. To say this clearly,

16. Isaac Newton, *Mathematical Principles of Natural Philosophy* (1687), quoted in *Science: Men, Methods, Goals,* ed. Baruch A. Brody and Nicholas Capaldi (New York: W. A. Benjamin, 1968), 78.

17. See David Bloor, *Knowledge and Social Imagery* (London: Routledge & Kegan Paul, 1977). This sociology of science is flawed, as I explain later, but illuminating nevertheless.

18. Bhaskar, *Reclaiming Reality,* points out that the astounding flexibility and adaptiveness of this theory of science is dependent on its adherents' refusal to acknowledge that there is any theory of science in play at all.

however, is to challenge directly and deeply the positivist grounds of the sciences' excessively empiricist theory of science.[19] Either scientific method in fact leaves a great deal to the "wit and imagination," or else scientific method should be taken to include processes of deciding how we should shape the entire moral and political order. In Millman and Kanter's passage there is, as I have noted, an unselfconsciousness about the paradoxical way the parts of their statement fit together. Is it really an "unspoiled vision" that the women's movement brings, not one shaped by interests in improving women's condition? Are there no interests, no desired benefits to men as men—whether or not consciously intended—that might account for the "covers and blinders" over women's as well as men's eyes? Did we all really "persuade ourselves" of the truth of the partial and distorted sexist vision of the world? Why haven't our lives improved as rapidly as the story about the townspeople would predict? Women's movements have been removing covers and blinders from eyes in the West at least since Christine de Pisan wrote *The City of Ladies* in the fifteenth century, yet we still live in a world ruled by powerful old naked patriarchal emperors.

I make these comments not as a criticism of Millman and Kanter, for I think I would have said much the same thing to the audience they were addressing. (In fact, I hear criticisms of bad science not very different from theirs emerging from my very own lips when I am initially presenting feminist materials to audiences that I judge to be friendly to conventional views of science but hostile to what they consider feminism.) Rather, I make these points to indicate how difficult it is for feminists to maintain a theory of science that is coherent with the sciences' own visions of themselves. The attempt to explain within such constraints the resources that the women's movement generates for the growth of scientific knowledge reveals the flaws in the paternal discourse.

The critics of bad science appear to be caught between two loyalties. On the one hand, they try to respect the dogma that one can explain "good science" without referring to its social causes. On the other hand, they think that the women's movement *is* a social cause of better science and that an understanding of *why* it is should inform scientific practice at least to the extent that scientists should welcome the wom-

19. A reminder to colleagues in the natural sciences: what is at issue here is a philosophy, a theory, of the history and practice of science known as empiricism, not the desirability of empirical research.

en's movement and listen to what it says in order to increase the growth of knowledge.

Criticisms of Science-as-Usual

How Women Scientists Benefit Science

The logic of the Marxist theory of science, from which the criticisms of science-as-usual borrow, does not lead to advocating the advancement of women in the existing scientific enterprise if that means leaving science otherwise unchanged. From this perspective, at best it makes no difference at all to women's situation in general if women are added to the social structure of a science that appears to be so thoroughly integrated with the misogynist, racist, and bourgeois aspects of the larger society. More likely it is a bad thing, since it diverts women's attention and energies from struggles against the *sources* of male domination and adds their energies to science's misogynist, racist, and bourgeois tendencies (whether or not these are intended by individual scientists).

Moreover, adding women to an institution that is highly stratified by class and race as well as by gender has the effect of strengthening class and race divisions between women. Women at the top of race and class hierarchies who succeed in science tend not to criticize or work against those forms of domination that oppress their sisters in other classes and races; they can easily become mere tokens whose individual achievement has little or no positive effect on the situation of the women who are not so favored. This is not to say that these women have not had to struggle mightily and unfairly to achieve the credentials and positions that flow so much more routinely to their male colleagues, nor is it to say that they intend such consequences. Nevertheless, it is frequently the case that their hard-won success does not significantly improve the situation for other women. It may sometimes even prove detrimental: if hiring or promoting a few (compliant) members of a protected class satisfies the watchdogs of affirmative action, the hiring of other highly qualified members of those classes can be more effectively resisted. Successful (and unsuccessful) women who say "I've never experienced sexism" invariably have done nothing to challenge what was expected of them as women. They have not taken the risks that others have taken to open the doors through which they now walk.[20] From the perspective of the critics of science-as-usual, women scientists—intentionally or not— are in this way complicitous with male domination. On balance, adding women to science—if that is the only change one intends to make in science—strengthens an institution that should be weakened.

The Marxist theory of science argued that if its bourgeois shell could be stripped away, a purportedly pure science could emerge that would be useful to the working class. The feminist critics of science-as-usual borrow something of this notion. From this perspective, women *do* or at least *could* provide special resources, as women, to science. This is so not because women have some inherent and universal ways of reasoning, attributable either to their different biology or to "women's intuition."[21] Instead, it is so because of the gap between women's experiences and the dominant conceptual schemes, from which have emerged so many issues of the women's movement and the most important feminist research in social science and biology. All the issues about women's bodies arise from this gap: issues of reproduction, child care, the assignment to women of the care of everyone's bodies and of the local places where they exist, sexuality, rape, incest, wife-battering, the mutilation caused by standards of beauty.[22] So too does the focus on gender itself: gender appears, emerges, as a phenomenon we can all see only from the perspective of women's lives. This gap is also the source of all the criticisms of the exclusion from and distortion of women and their lives in dominant patterns of Western thought— including scientific thought.

In this context, the "bifurcated consciousness" of the alienated woman sociologist is a great resource for the radical transformation of sociology, as Dorothy Smith argues. I think that we can generalize her argument to one about women scientists and the radical transformation of the sciences. The conceptual scheme of male scientists matches far too comfortably the dominant concepts of ruling; their sciences help to produce the conceptual forms for ruling in our kind of social

20. Rossiter, *Women Scientists in America,* tells the stories of these struggles in the United States.

21. This is a common and convenient way to misread the feminist standpoint theorists and the object relations psychology that has informed some of their accounts. Parts II and III discuss this misreading further.

22. Smith, *Everyday World.*

order. This is in large part what the feminist critics of science's sexual metaphors are arguing: men unselfconsciously find it natural to think in terms of "the scientist, he . . ." and "nature, she . . ." Metaphors of control and the domination of nature feel more fruitful to men in the dominant groups than to women as ways to think about nature. Men—but rarely women—have for centuries projected their fears and desires onto the canvas of the world around them through the scientific pictures they construct.[23]

Consequently, the unexamined conceptual world of male scientists is an impoverished resource with which to try to explain *critically* our social order. It is as if an anthropologist were to restrict herself to the conceptual scheme of the indigenous people in attempting to explain aspects of their daily life, though because she has not been socialized into the culture's way of thinking, she can detect aspects of their beliefs and behaviors that are invisible to them. Smith points out that as both a woman and a sociologist (or, we can infer, a scientist more generally), one has access to firsthand experience of the kinds of women's daily activities that make it possible for other people not to have to think about their bodies, or the local places where those bodies exist, and also to the activities of the "ruling gender." With the help of theory, one can come to understand the relations between the two kinds of activities—how the former make the latter possible, and how the latter structure the way the former will occur. Similar analyses of the important contributions that can be made to the growth of knowledge by looking at the world from the gap between women's and men's activities have been provided by other thinkers cited above.[24]

The focus of these arguments has been the social sciences, human biology, and evolutionary theory, and those are fields where one would expect them to be most fruitful. It is not impossible, however, to produce this kind of account in the natural sciences as well. Historians of

23. See Evelyn Keller, *Reflections on Gender and Science* (New Haven, Conn.: Yale University Press, 1984); Carolyn Merchant, *The Death of Nature: Women, Ecology, and the Scientific Revolution* (New York: Harper & Row, 1980); Schiebinger, *The Mind Has No Sex.*

24. Biologists whose arguments for a feminist science express important aspects of this approach to science and knowledge include Lynda Birke, *Women, Feminism, and Biology* (New York: Methuen, 1986); Ruth Bleier, *Science and Gender: A Critique of Biology and Its Theories on Women* (New York: Pergamon Press, 1984), chap. 8; Anne Fausto-Sterling, *Myths of Gender: Biological Theories about Women and Men* (New York: Basic Books, 1985), chap. 7.

science and of theories of scientific knowledge such as Susan Bordo, Evelyn Fox Keller, Carolyn Merchant, Donna Haraway, Hilary Rose, and Londa Schiebinger have shown that extremely abstract elements of Western scientific thought gained legitimacy because they both reflected and reinforced certain historical aspects of the experiences of men in the dominant groups.[25] From the perspective of the standpoint theorists' criticisms of science-as-usual, then, women scientists can bring certain benefits to the growth of knowledge if they can find ways to use their experience as women, informed by feminist theorizing, to create a critical perspective on the dominant conceptual schemes and how they shape scientific research and practice. (I am not suggesting that the historians are all critics of science-as-usual, as I have been describing this position, but that their analyses permit such accounts of the natural sciences.)

How can we justify a feminist world that has no women scientists? Should we imagine creating sciences *for* women that are not made *by* women (as well as by men)? I hasten to mention that standpoint theorists do not argue that all women scientists automatically bring such benefits to science, or that only women can look at the world from the perspective of women's activities. They argue only that powerful critical theories *can be developed* out of the gap between the perspective from that part of human activity that is assigned to women and the conceptual schemes grounded only in the ruling part, which men of the dominant classes and races reserve for themselves.

Thus the feminist critique of science-as-usual provides a different assessment of the resources for science that women scientists can provide. Whereas the critics of bad science think it is primarily their numbers and their genderfree talents and abilities that make an important contribution to the growth of knowledge, what is important to critics of science-as-usual is their ability to think from the perspective of the social activities assigned to women. I argue that men too can learn to think from the perspective of women's activities; perhaps John Stuart Mill, Marx and Engels, Frederick Douglass, and other male feminists have done so (see Chapter 11). But men can do so only after women have articulated what the gap is between their experience and

---

25. Susan Bordo, *The Flight to Objectivity* (Albany: State University of New York Press, 1987); Keller, *Reflections;* Merchant, *Death of Nature;* Donna Haraway, *Primate Visions: Gender, Race, and Nature in the World of Modern Science* (New York: Routledge, 1989); Rose, "Hand, Brain, and Heart"; Schiebinger, *The Mind Has No Sex.*

the dominant conceptual schemes, and after men too have engaged in the political struggles described next.

How Feminist Politics Benefits Science

Most critics of science-as-usual hold that political struggle is a necessary part of learning how to criticize the dominant conceptual schemes from the standpoint of women's activities. Political struggle is overtly "inside science" for this approach—not out there somewhere in the environment for our mental appreciation. "His resistance is the measure of your oppression," a saying from the early 1970s, suggested that only through political struggle could women get the chance to observe the depth and extent of masculine privilege. And feminist struggle has direct scientific value. As Nancy Hartsock explains:

> Because the ruling group controls the means of . . . [all] production, the standpoint of the oppressed represents an achievement both of science (analysis) and of political struggle on the basis of which this analysis can be conducted. . . . Women's lives, like men's, are structured by social relations which manifest the experience of the dominant gender and class. The ability to go beneath the surface of appearances to reveal the real but concealed social relations requires both theoretical and political activity. Feminist theorists must demand that feminist theorizing be grounded in women's material activity and must as well be a part of the political struggle necessary to develop areas of social life modeled on this activity.[26]

To put this point another way, science is like sculpture in that it is a "craft" activity. Only working with (and against) the material reveals its true character—its internal relations and structure; the deepest, most enduring, and most powerful sources of its strength; its surprising weaknesses. The "material" of which feminist politics can reveal the regularities and underlying causal tendencies consists of the gender relations that are again and again, in different historical forms, part of the politics within which sciences are constructed and reconstructed.

This claim can call on historical precedents. The feminist politics of the women's movement provides the kinds of resources to science that were provided for the emergence of modern science by the political struggles necessary to bring Europe from the medieval world to moder-

26. Hartsock, "The Feminist Standpoint," 288, 304.

nity. That is, modern science itself was created through a movement of social liberation. The new physics advanced precisely because it both expressed the ethos of an emerging class (materialism, antielitism, progress) and also provided the means for expressing that ethos in technologies that could materially advance that class. Its very "method"—experimental observation—required the performance of both head and hand labor by one and the same person, by new kinds of persons who did not exist in the feudal aristocracy.[27] The early modern scientists were engaged in these political struggles. Are not women sociologists and other women scientists just such new kinds of persons, created through the politics of a social liberation movement?

This is a robust analysis of the positive role that feminist politics plays in the growth of scientific knowledge. Feminist struggle is a fundamental part of gaining knowledge, including knowledge about and through science. People, men as well as women, who do not engage in it, who do not risk in their daily activities offending or threatening the legitimacy of male supremacy in any of its encultured forms, cannot *know* how the social and natural worlds are organized. As I mentioned earlier, a woman who can say "I've never been discriminating against as a woman" has not engaged in those political struggles in personal, community, or institutional contexts which patriarchy finds so threatening. Engaging in those struggles is an activity that is inside—part of—science for the critics of science-as-usual.

Feminist Theory about Science as a Scientific Resource

Finally, for the critics of science-as-usual, good theory about science is crucial to doing good science. What happens in the lab begins far away, out in the moral, economic, and political activities of the society. The absence of an explicit theory about the causes of the patterns of everyday life in the labs or the sociology department results in making scientists simply "guns for hire," myopically pursuing the production of information—for whom and for what purposes they haven't the slightest idea and think that they are not supposed to care.

This argument has been made forcefully by feminist biologists and

27. Merchant, *Death of Nature;* Van den Daele, "Social Construction of Science"; Edgar Zilsel, "The Sociological Roots of Science," *American Journal of Sociology* 47 (1942).

critics of the sexist production of scientific technologies. They point out that one needs an adequate theory about science in order to begin to eliminate the ways in which science and its technologies victimize women. The problem is not that there are sexist and androcentric "misuses and abuses" of scientific technologies such that they could be ended and leave pure, gender-impartial sciences and technologies. Instead, it is inevitable that women will be victimized by the sciences and their technologies in a society such as ours where women have little power, where almost all scientific research is technology-driven, and where political issues are posed as requiring merely technological "solutions."

Within the natural sciences, however, there is immense resistance to social theory about science (the reasons are discussed in the next chapter). Natural scientists are trained to think that they should be their own uniquely legitimate experts on what their work is "about," and that what it is about is entirely contained within the consciousness of the scientists. From the perspective of the critics of science-as-usual, because the social processes that eventuate in everyday life in the laboratories begin far away from the laboratories—in a sluggish economy, in the desire to win a war in Asia, in attempts to control African Americans in cities, in desires to limit population growth in Asia—firsthand expertise in the laboratories won't go far toward providing an explanation of laboratory life. Evelyn Keller, trained as a physicist and then as a molecular biologist, illuminates the struggle that enabled her to understand that the natural sciences were not necessarily always the best place to start, even in our most rational efforts to gain knowledge.

> By paying attention to my own unwitting thoughts, words, and actions (and in the process coming to see their internal coherence), I became a direct witness of the force of beliefs—particularly of beliefs that are not conscious. . . . No longer did I take natural science as the necessary starting point for inquiry about the nature of the world (including the human race). Once having seen the place and force of beliefs and feelings in even our most rational endeavors, it became possible to make the analysis of our own subjectivity a starting point for an inquiry into the nature of science.[28]

28. Evelyn Fox Keller, "Women Scientists and Feminist Critics of Science," *Daedalus* 116:4 (1987), 87–88.

She went on to provide one of the most compelling and illuminating of the recent feminist critiques of science.[29] Thus feminist theory about science must be seen as inside the process of science, where it can help scientists explain the social conditions in both scientific institutions and the surrounding society that tend to encourage empirically more adequate beliefs; identify background assumptions that tend to distort the results of research; conceptualize and design research in ways that avoid powerful cultural biases; interpret and select data to produce the most reliable evidence for and against hypotheses. In short, feminist theory can help scientists learn all the things good scientists want to know.

In later chapters I develop this point in another way to argue that we should think of the natural sciences as being *inside* critical social sciences because the objects-of-knowledge—"nature, herself"—never come to science denuded of the social origins, interests, values, and consequences of their earlier "careers" in social thought. No matter how rigorously science attempts to "strip nature bare" and make her "reveal her secrets," there are always more "veils" to be found. Scientific inquiry itself, it turns out, is busy continuously weaving those veils. Nature-as-object-of-knowledge is more than a cultural construct, but it is always that.

I have explored two views of what resources the women's movement can provide for science. It seems clear that feminists who hold these views have different audiences for their feminisms. Each of us may well find one set of beliefs more congenial than the other to our own projects and the institutional worlds in which we carry them out. Historically, we have too often tended to express these affinities in terms that devalued the projects of the "other feminism." I think it would be a great loss for the women's movement to abandon either approach.

The critics of "bad science" are most active in supporting efforts to get more women into science—partly because these critics are most often to be found in the sciences. Other feminists may think that we would all be better off if the women now in the sciences devoted their prodigious talents and energies to achieving more obvious or direct benefits to women. But though the benefits of having more women in

29. Keller, *Reflections*.

the sciences may not in themselves seem great to feminists not interested in these fields, there are important reasons to want more *feminists* in science, and—with some exceptions—women do seem to turn into feminists more quickly and thoroughly than do men. So why not encourage the entrance into this institution of the social group most likely to produce feminists?

We need feminists in the existing sciences for many reasons: to blow the whistle from within on the failures of scientists to adhere to their often expressed principles of impartiality, disinterest, value-neutrality; to draw into their agendas "prefeminists" in the sciences (male and female) who are open to their criticisms of science; to gain for women access to the status and authority in the larger society that such positions bring; where possible, to explain what women need to know about the regularities and underlying causal tendencies of nature and social life; to generate, within equal opportunity justifications, scientific projects that are specifically in women's interests. (These purposes do not appear to me to be different from the feminist agendas of women who work within other powerful institutions in our society— the government, for instance, or higher education. A decision to work within any such institution should not mean that one has made a commitment to work on every project that might be generated there; blind obedience to superiors is not supposed to be desired by superiors in these institutions.) It becomes an important goal for feminists within the sciences to help locate the trail of social relations connecting what occurs in the laboratories with the social relations of the larger society and to contribute to the demystification of science, to the increase of scientific literacy, to the generation of projects—strategic as well as substantive that can benefit women.

The women's movement also needs feminists *outside* science to undertake a critical examination of the regularities and underlying causal tendencies in the fit of science with other social institutions. (By "outside science" I mean only outside the labs or their equivalent; as I have argued, there is virtually no "outside" to science in the contemporary West.) The study of science is itself a science, and people inside a culture or institution are never in the best position to see those causes of their daily activities that originate far outside their daily world. These understandings can have an effect on how people inside science think about their activity and make intellectual, practical, and political choices as scientists. But to have this effect, feminists inside and outside the sciences need to think of themselves as working together in spite of the occasionally contradictory aspects of their conjoined efforts.

To paraphrase a metaphor familiar to philosophers, the women's movement finds itself in the middle of the ocean, having both to re-design and rebuild the leaky ship of modern science one plank at a time. (Of course, it finds resources in other liberatory science movements, as I later discuss.) In other words, we are forced to remake science but not under conditions of our own choosing.

4

Why "Physics" Is a Bad Model for Physics

Both natural and social sciences can benefit from feminism in the variety of ways preceding chapters have described. Even most feminist critiques, however, have not gone far enough in identifying the fortifications that have been erected—intentionally or not—around the natural sciences and that protect them from the very kind of critical, causal scientific explanation that the natural sciences insist on for all other social phenomena. This chapter focuses on popular but false beliefs that block our ability to understand the natural sciences as a social phenomenon and, consequently, to appreciate the relevance of feminism to the content and logic of research and explanation.

Science without the Elephants

Are feminist criticisms of Western thought relevant to the natural sciences? "Of course, there should be more women in science, mathematics, and engineering—and the good ones will rise to the top," the conventional argument says. "Moreover, it is not at all good that some technologies and applications of natural science have been dangerous to women; policymakers should take steps to eliminate these misuses and abuses of the sciences. But the logic of research design and the logic of explanation in the physical sciences are fundamentally untouched by the feminist criticisms and will necessarily remain so. This is because the logic of research and of explanation and the cognitive, intellectual content of natural science's claims—'pure science'—cannot be influenced by gender."

This argument will not stand up to scrutiny. It is grounded not only in an underestimation of the pervasiveness of gender relations—relations that appear not only between individuals but also as properties of institutional structures and of symbolic systems[1]—but also in false beliefs about the natural sciences. Because of these beliefs, it is difficult to make sense of many aspects of science and society. One can think of these false beliefs as extraneous elements in metatheories of science: if we remove them, we can begin to understand aspects of science that appear inconsistent or inexplicable as long as we hold them.

By "physics"—in quotation marks—I mean a certain image of science that is full of these mystifying beliefs. "Physics" is magical; it is like the ancient image of a column of elephants holding up the earth. The logic of the column of elephants—"You can't fool me, young man: it's elephants all the way down," as the punch line to the old joke goes—prevents the observer from asking questions that would quickly come to mind were the elephants not so solidly in view. Physics is to "physics" as a satellite photo of the earth is to a picture of the earth balanced on top of a column of elephants. We can understand physics without "physics."

The reader should be reassured again that I do not intend to throw out the baby of science along with the bath water of false views about science.[2] My concern is to separate the false beliefs from those that are conducive to empirically, theoretically, and politically more adequate sciences—to identify more carefully where the baby ends and the bathwater begins. There are some *causes* of scientific beliefs and practices that are to be found outside the consciousnesses of individual scientists; that is, they are not *reasons* for the acceptance or rejection of these beliefs and practices. Our society is permeated by forms of scientific rationality; and it is in just such a society that there is a deep resistance to understanding how the institutional practices of science shape the activities and consciousnesses of scientists as well as of the rest of us. From the perspective of the democratic tendencies within

1. See Sandra Harding, *The Science Question in Feminism* (Ithaca: Cornell University Press, 1986), 52–56, for discussion of these three sometimes conflicting manifestations of gender relations.

2. Contrary to the apparent recommendation of such critics as, e.g., Sal Restivo, "Modern Science as a Social Problem," *Social Problems* 35:3 (1988).

science, that resistance is irrational, but it frames discussions in such a way that it is difficult for people to understand their own activities and why some of the choices they confront are so limited and narrow. The false beliefs examined below serve to hide the irrationality from critical scrutiny.

Some readers will think I am criticizing a straw figure. They will find it convenient to see only positivist tendencies that are no longer fashionable as the reasonable target of these criticisms. I cannot here detour to define positivism and debate its influence. But it is widely recognized in the social studies of science that although fewer scientists, philosophers, and social scientists who model their work on the natural sciences are as openly enthusiastic about positivism than was the case forty and more years ago, most of these people still happily embrace fundamental assumptions of positivism. As philosopher Roy Bhaskar has astutely observed, positivism still represents the unreflective "consciousness of science."[3]

Six False Beliefs

(1) "Feminism is about people and society: the natural sciences are about neither; hence, feminism can have no relevance to the logic or content of the natural sciences." One line of thinking behind this argument is that researchers are far more likely to import their social values into studies of other humans than into the study of stars, rocks, rats, or trees. And it is absurd, the conventionalist will argue, to imagine that social values could remain undetected in studies of the abstract laws that govern the movements of the physical universe. Scientific method has been constructed exactly to permit the identification and elimination of social values in the natural sciences. Practicing scientists and engineers often think the discussions of objectivity and method by philosophers and other nonscientists are simply beside the point. If bridges stand and the television set works, then the sciences that produced them must be objective and value-free—that's all there is to the matter.

One could begin to respond by pointing out that evolutionary theo-

---

3. Roy Bhaskar, "Philosophies as Ideologies of Science: A Contribution to the Critique of Positivism," in his *Reclaiming Reality* (New York: Verso, 1989).

ry, a theory that is about all biological species and not just about humans, clearly "discovered" secular values in nature, as the creationists have argued. It also "discovered" bourgeois, Western, and androcentric values, as many critics have pointed out.[4] Moreover, the physics and astronomy of Newton and Galileo, no less than those of Aristotle and Ptolemy, were permeated with social values. Many writers have identified the distinctively Western and bourgeois character of the modern scientific world view.[5] Some critics have detected social values in contemporary studies of slime mold and even in the abstractions of relativity theory and formal semantics.[6] Conventionalists respond by digging in their heels. They insist on a sharp divide between premodern and modern sciences, claiming that while medieval astronomy and physics were deeply permeated with the political and social values of the day, the new astronomy and physics were (and are) not; this is exactly what distinguishes modern science from its forerunners. As historian of science Thomas Kuhn said, back when he was such a conventionalist, the world view characteristic of medieval Europe was much like that of "primitive societies" and children, which "tends to be animistic. That is, children and many primitive peoples do not draw the same hard and fast distinction that we do between organic and inorganic nature, between living and lifeless things. The organic realm

4. Stephen Jay Gould, *The Mismeasure of Man* (New York: Norton, 1981); Ruth Hubbard, "Have Only Men Evolved?" in *Discovering Reality: Feminist Perspectives on Epistemology, Metaphysics, Methodology, and Philosophy of Science,* ed. Sandra Harding and Merrill Hintikka (Dordrecht: Reidel, 1983).

5. See Leszek Kolakowski, *The Alienation of Reason: A History of Positivist Thought* (New York: Doubleday, 1968); Carolyn Merchant, *The Death of Nature: Women, Ecology, and the Scientific Revolution* (New York: Harper & Row, 1980); Alfred Sohn-Rethel, *Intellectual and Manual Labor* (London: Macmillan, 1978); Margaret C. Jacob, *The Cultural Meaning of the Scientific Revolution* (New York: Knopf, 1988); Wolfgang Van den Daele, "The Social Construction of Science," in *The Social Production of Scientific Knowledge,* ed. Everett Mendelsohn, Peter Weingart, and Richard Whitley (Dordrecht: Reidel, 1977); Morris Berman, *The Reenchantment of the World* (Ithaca: Cornell University Press, 1981).

6. Evelyn Fox Keller, "The Force of the Pacemaker Concept in Theories of Aggregation in Cellular Slime Mold," and "Cognitive Repression in Contemporary Physics," both in Keller, *Reflections on Gender and Science* (New Haven, Conn.: Yale University Press, 1985); Paul Forman, "Weimar Culture, Causality, and Quantum Theory, 1918–1927: Adaptation by German Physicists and Mathematicians to a Hostile Intellectual Environment," *Historical Studies in the Physical Sciences* 3 (1971); Merrill B. Hintikka and Jaakko Hintikka, "How Can Language Be Sexist?" in Harding and Hintikka, *Discovering Reality.*

has a conceptual priority, and the behavior of clouds, fire, and stones tends to be explained in terms of the internal drives and desires that move men and, presumably, animals."[7]

The conventionalist fails to grasp that modern science has been constructed by and within power relations in society, not apart from them.[8] The issue is not how one scientist or another used or abused social power in doing his science but rather where the sciences and their agendas, concepts, and consequences have been located within particular currents of politics. How have their ideas and practices advanced some groups at the expense of others? Can sciences that avoid such issues understand the causes of their present practices, of the changing character of the tendencies they seem to "discover in nature" in different historical settings?

Even though there are no complete, whole humans visible as overt objects of study in astronomy, physics, and chemistry, one cannot assume that no social values, no human hopes and aspirations, are present in human thought about nature. Consequently, feminism can have important points to make about how gender relations have shaped the origins, the problematics, the decisions about what to count as evidence, social meanings of nature and inquiry, and consequences of scientific activity. In short, we could begin to understand better how social projects can shape the results of research in the natural sciences if we gave up the false belief that because of their nonhuman subject matter the natural sciences can produce impartial, disinterested, value-neutral accounts of a nature completely separate from human history.

(2) "Feminist critics claim that a social movement can be responsible for generating empirically more adequate beliefs about the natural world. But only false beliefs have social causes. Whatever relevance such critics have to pointing out the social causes of false beliefs, feminism can not generate 'true beliefs.'" This claim assumes that no social science findings could be relevant to our explanations of how the best, the empirically most supported (or least refuted) hypotheses arise and gain scientific legitimacy. Some conventionalists will agree that the

7. Thomas Kuhn, *The Copernican Revolution* (Cambridge, Mass.: Harvard University Press, 1957), 96.

8. See Joseph Rouse, *Knowledge as Power: Toward a Political Philosophy of Science* (Ithaca: Cornell University Press, 1987); Merchant, *Death of Nature;* Van den Daele, "Social Construction of Science"; Harding, *The Science Question in Feminism,* chaps. 8–9.

social sciences can tell us about the intrusion of social interests and values into research processes that have produced false beliefs: when we want to know why phlogiston theory, phrenology, Nazi science, Lysenkoism, and creationism were able to gain a legitimacy and respect that they should not have had, the causes are to be found in social life. Funding them is a worthy task for sociologists and historians. But the content of "good science" has no social causes, only natural ones, according to the conventionalist. It is a result of the way the world is, of the way our powers of observation and reason are, and of bringing our powers of observation and reason to bear on the way the world is. Consequently, the most widely accepted natural science claims require no causal accounts beyond the reason scientists could give for their own cognitive choices.

Supporting this view of the asymmetry of causal explanations of belief is a long tradition in epistemology but one that has been criticized in recent decades by sociologists of knowledge.[9] They argue that it is simply a prejudice of philosophers to hold that the beliefs a culture regards as legitimate should uniquely be excepted from causal social explanations. To hold such a position is to engage in mysticism; it is to hold that the production of scientific belief, alone of all distinctively human social activities, has no social causes. Instead, they argue, a fully scientific account of belief will seek causal symmetry; it will try to identify the social causes (as well as the natural ones) of the best as well as the worst beliefs.

This sociological account is flawed in a variety of ways. For one thing, these writers appear to exempt their own claims from the causal accounts they call for elsewhere, in this and other ways adopting still excessively positivist conceptions of scientific inquiry.[10] Moreover, their account appears to reduce scientific claims to beliefs that happen to be socially acceptable. It offers no way to talk about the natural constraints within which historically distinctive scientific accounts are

9. David Bloor, *Knowledge and Social Imagery* (London: Routledge & Kegan Paul, 1977); Barry Barnes, *Interests and the Growth of Knowledge* (Boston: Routledge & Kegan Paul, 1977); Karin Knorr-Cetina, *The Manufacture of Knowledge* (Oxford: Pergamon Press, 1981); Karin Knorr-Cetina and Michael Mulkay, eds., *Science Observed: Perspectives on the Social Study of Science* (Beverly Hills, Calif.: Sage, 1983).

10. See, e.g., Bloor, *Knowledge and Social Imagery*, 142–44. Attempts to remedy this situation by pursuing to its amusing though disastrous end the embrace of relativism required by the logic of the "strong programme" in the sociology of knowledge can be seen in Steve Woolgar, ed., *Knowledge and Reflexivity* (Beverly Hills, Calif.: Sage, 1988).

produced.[11] But we do not have to replicate the limitations of these sociological accounts, the functionalism and relativism that plagues these otherwise illuminating analyses. We can hold that our own (true! or, at least, less false) account also has social causes—that, for example, changes in social relations have made possible the emergence of the distinctive intellectual and political trajectory of modern science as well as of feminism. These histories leave their fingerprints on the cognitive content of science no less than of feminism.[12] Moreover, we can insist that the identification of social causes for the acceptance of a belief does not exclude the possibility that that belief *does* match the world in better ways than its competitors. That is, we can hold that certain social conditions make it possible for humans to produce reliable explanations of patterns in nature, just as other social conditions make it very difficult to do so.

If the objection to feminist accounts of the social causes of "true belief" were reasonable, one would have to criticize on identical grounds the new histories, sociologies, psychologies, anthropologies, and political economies of science. A wide array of studies have shown the politics within which modern scientific knowledge has been constructed. Eliminating the idea that only false beliefs can have social causes—this "elephant"—makes possible more coherent accounts of what actually has contributed to the growth of knowledge in the history of the sciences. It makes possible an understanding of feminism as able to advance knowledge not only by debunking false beliefs but also by helping to create social conditions conducive to the recognition of less partial and distorting beliefs, and by generating such scientifically preferable beliefs.

(3) "Science fundamentally consists only of the formal and quantitative statements that express the results of research, and/or science is a unique method. If feminists do not have alternatives to logic and mathematics or to science's unique method, then their criticisms may be relevant to sociological issues but not to science itself." Galileo

11. Hilary Rose, "Hyper-reflexivity: A New Danger for the Counter Movements," in *Counter-Movements in the Sciences: The Sociology of the Alternatives to Big Science,* ed. Helga Nowotny and Hilary Rose (Dordrecht: Reidel, 1979).

12. See, e.g., Harding, "Why Has the Sex/Gender System Become Visible Only Now?" in Harding and Hintikka, *Discovering Reality;* Van den Daele, "Social Construction of Science"; and Edgar Zilsel, "The Sociological Roots of Science," *American Journal of Sociology* 47 (1942).

argued that nature speaks in the language of mathematics, so if we want to understand nature, we must learn to speak "her" language. Some conventionalists have understood this to mean that "real science" consists only of the formal statements that express such laws of nature as those discovered by Isaac Newton, Robert Boyle, and Albert Einstein.

There can appear to be no social values in results of research that are expressed in formal symbols; however, formalization does not guarantee the absence of social values.[13] For one thing, historians have argued that the history of mathematics and logic is not merely an external history about who discovered what when. They claim that the general social interests and preoccupations of a culture can appear in the forms of quantification and logic that its mathematics uses. Distinguished mathematicians have concluded that the ultimate test of the adequacy of mathematics is a pragmatic one: does it work to do what it was intended to do?[14]

Moreover, formal statements require interpretation in order to be meaningful. The results of scientific inquiry can count as results only if scientists can understand what they refer to and mean. Without decisions about their referents and meanings, they cannot be used to make predictions, for example, or to stimulate future research. And as is the case with social laws, the referents and meanings of the laws of science are continually extended and contracted through decisions about the circumstances in which they should be considered to apply.

There is also the fact that metaphors have played an important role in modeling nature and specifying the appropriate domain of a theory.[15] To take a classic example, "nature is a machine" was not just a useful heuristic for explaining the new Newtonian physics but an inseparable part of that theory, one that created the metaphysics of the theory and showed scientists how to extend and develop it. Thus, social metaphors provided part of the evidence for the claims of the new sciences; some of their more formal properties still appear as the kinds of relations model-

---

13. This section repeats some of the arguments made in Harding, *The Science Question in Feminism*, chap. 2.

14. Bloor, *Knowledge and Social Imagery;* Morris Kline, *Mathematics: The Loss of Certainty* (New York: Oxford University Press, 1980).

15. Mary Hesse, *Models and Analogies in Science* (Notre Dame, Ind.: University of Notre Dame Press, 1966); Merchant, *Death of Nature.* See also my discussion of Hesse's conclusions in Harding, *The Science Question in Feminism*, 233–39.

ed by the mathematical expressions of the natural sciences. They were not only "outside" the process of testing hypotheses; they were also "inside" it. The social relations of the period, which both made possible and were in turn supported by the machines on which Newton's mechanistic laws were modeled, functioned as—*were*—part of the evidence for Newtonian physics. Giving up the belief that science is really or fundamentally only a collection of mathematical statements is necessary if we are to begin to explain the history and practices of science. Insistence on this belief is a way of irrationally restricting thought.

If science is not reducible to its formal statements, is it reducible to its method? This is an equally problematic claim. Contemporary physicists, ethologists, and geologists collect evidence for or against hypotheses in ways different from those that medieval priests used to collect evidence for or against theological claims, yet it is difficult to identify or state in any formal way just what it is that is unique about the scientific methods. For one thing, different sciences develop different ways of producing evidence, and there is no clear way to specify what is common to the methods of high-energy physics, ethology, and plate techtonics. "Observing nature" is certainly far too general to specify uniquely scientific modes of collecting evidence; gatherers and hunters, premodern farmers, ancient seafarers, and mothers all must "observe nature" carefully and continuously in order to do their work. These examples also show that linking prediction and control to the observation of nature are certainly not unique to science, since they are also crucial to gathering and hunting, farming, navigation, and child care. Scientific practices are common to every culture. Moreover, many phenomena of interest to science, though they can be predicted and explained, cannot be controlled—for example, the orbit of the sun and the location of fossils. And prediction alone is possible on the basis of correlations that in themselves have little or no explanatory value.

Philosophers and other observers of science have argued for centuries over whether deduction or induction should be regarded as primarily responsible for the great moments in the history of science,[16] but it is obvious that neither is unique to modern science: infants and dogs regularly use both. It may be futile to try to identify distinctive features of knowledge-seeking that will exclude mothers, cooks, or

---

16. See Sandra Harding, ed., *Can Theories Be Refuted? Essays on the Duhem-Quine Thesis* (Dordrecht: Reidel, 1976).

farmers from the ranks of people who should be counted as scientists but will include highly trained but junior members of, say, biochemical research teams. This is even more true in a society such as ours where scientific rationality has permeated child care, cooking, and farming.

One might try to defend the idea that the important feature of scientific method is science's critical attitude.[17] That is, scientific method is fundamentally a psychological stance. In all other kinds of knowledge-seeking, this line of argument goes, we can identify assumptions that are regarded as sacred or immune from refutation; only modern science holds all its beliefs open to refutation. But this proposal is not supported by the history, present practices, or leading contemporary metatheories of science. On the one hand, assumptions that are held immune from criticism—either on principle or inadvertently—are never absent from the sciences. The history of science shows that scientists and science communities again and again make unjustified assumptions and that they are loath to examine critically the hypotheses in whose plausibility they have invested considerable time, energy, and reputation. Moreover, we could call some beliefs constitutive of science in the sense that they can be questioned only at the risk of creating skepticism about the whole scientific enterprise. One example is the idea that all physical events and processes have causes even if we can't always know what they are; another is that it is a good thing to know more about nature. Furthermore, everyone understands that there must be many scientific assumptions that are questionable in principle, but that they cannot all be questioned simultaneously if research is to occur at all. Thomas Kuhn proposed that a field of inquiry really becomes a science only when it decides to accept some set of beliefs as "not to be contested" and makes these the assumptions that define the field (this line of thought led Kuhn to dubious claims about how to create true sciences, as we shall see).[18] Others point to the necessarily unquestioned "background assumptions" or "auxiliary hypotheses" that inevitably hover behind every hypothesis being tested. These include optical theories, beliefs about how the testing and recording instruments work, assumptions about which variables are significant and about what can count as a repeated observation or experiment.

17. Robin Horton, "African Traditional Thought and Western Science," pts. 1–2, *Africa* 37 (1967); Karl Popper, *Conjectures and Refutations: The Growth of Scientific Knowledge*, 4th ed. (London: Routledge & Kegan Paul, 1972).

18. Thomas Kuhn, *The Structure of Scientific Revolutions* (Chicago: University of Chicago Press, 1970).

Nor is Western science the only domain of critical thought. All of us must have a critical attitude toward a good number of beliefs if we are to survive the vicissitudes of nature and social life. It is part of the ethnocentrism of the West to assume that only practitioners of Western scientific rationality exercise critical reason. Feminists and the working class have also questioned the assumption that critical reason is the talent only of the dominant groups.[19]

The idea that science really or fundamentally comprises formal statements or is a distinctive method is an extraneous belief that blocks our ability to describe and explain the workings of modern Western science. Science has many interlocking practices, products, referents, and meanings. It is a cumulative tradition of knowledge. It is an "origins story," a fundamental part of the way certain groups in the modern West identify themselves and distinguish themselves from others. It is a metaphysics, an epistemology, and an ethics. It is a politics that has been compatible with the agendas of modern liberal states, capitalism, and Protestantism. Some have pointed out not only that science has become a religion for many but that although it attempts to hide its religious character by distancing itself from religion, it intends to hold the place of a religion. What else, they ask, could one conclude about its insistence on its own absolute authority, on its "monologue" form, on its inherent moral good; about its intolerance of criticisms from "outside"; about its intended use to define the borders of "civilization"? It is a social institution with complex rituals and practices that both reflect and shape social relations in the cultures in which it exists. It is both the producer and the beneficiary of technological invention. It is a major factor in the maintenance and control of production and, increasingly, reproduction.

There is a striking contrast between this array of descriptions of "what science is" and the restricted range upon which conventionalists

19. See, e.g., Mary G. Belenky, B. M. Clinchy, N. R. Goldberger, and J. M. Tarule, *Women's Ways of Knowing: The Development of Self, Voice, and Mind* (New York: Basic Books, 1986); Carol Gilligan, *In a Different Voice: Psychological Theory and Women's Development* (Cambridge, Mass.: Harvard University Press, 1982); Sandra Harding, "Is Gender a Variable in Conceptions of Rationality? A Survey of Issues," in *Beyond Domination: New Perspectives on Women and Philosophy*, ed. Carol C. Gould (Totowa, N.J.: Littlefield, Adams, 1983); Genevieve Lloyd, *The Man of Reason: "Male" and "Female" in Western Philosophy* (Minneapolis: University of Minnesota Press, 1984); J. E. Wiredu, "How Not to Compare African Thought with Western Thought," in *African Philosophy: An Introduction*, 2d ed., ed. Richard A. Wright (Washington, D.C.: University Press of America, 1979).

insist. False beliefs block our ability to explain how science works.

(4) "Applications of science are not part of science proper. So feminist criticisms of the misuses and abuses of the sciences (such as of the proliferation of dangerous reproductive technologies) challenge only public policy about science, not science itself." Preceding discussions indicate why this statement is a distorted representation of science and technology and the relations between them. Whatever was true in the past, it is difficult now to identify anything at all that can count as *pure* science. Is this too strong a claim? Let us see. Science makes use of technological ideas and artifacts at least as much as the reverse. Moreover, even when scientific ideas do not result in any immediate application, they may very well still be permeated with values. After rethinking the complex relationship between sciences and technologies, many observers have concluded that science is "politics by other means." It is more than that, but it is that.

Everyone is willing to acknowledge that scientific research makes possible new technologies and applications of science. Science produces information that can be applied in the social world and used to design new technologies. This is not thought to threaten the purported purity of science, because it is not scientists but policymakers who actually decide to construct the technologies and carry out the new applications of scientific information. "You can't infer an 'ought' from an 'is,'" as philosophers like to say. Deciding what we ought to do with the information that science provides is supposed to be a separate process from producing the information in the first place. According to this way of thinking, it is policymakers who should be held responsible for the misuses and abuses of the sciences and their technologies—not scientists or the sciences themselves.

Because two distinct groups of people have responsibility for the two kinds of decisions, it is easier to think that technologies and sciences must be conceptually and politically separate. Scientists in universities and research laboratories produce the information; scientists in industry, the military, and the government make the decisions about what information is to be disseminated and how it is to be used.[20] But this division of labor does not have the consequences its defenders suppose.

---

20. See, e.g., Paul Forman, "Behind Quantum Electronics: National Security as Bases for Physical Research in the U.S., 1940–1960," *Historical Studies in Physical and Biological Sciences* 18 (1987).

It simply makes it difficult for scientists in universities to explain their own activities in a plausible way: that is, to give the kind of causal account of science that scientists recommend we give about everything else. Their explanations of their activities do not maximize coherence, generality, simplicity, do not fit with empirical evidence, and so on.

In the first place, some "is's" in practice ensure "oughts." For example, in a racist society, "pure descriptions" of racial difference have little chance of functioning as pure information. One can be confident that racist assumptions will markedly narrow the range of "reasonable" applications of such "information." Moreover, the very concern with racial difference in such a culture cannot be free of race value.[21] The scientific reports can be as value-neutral as possible in the sense that they describe only difference, not inferiority and superiority, and make no recommendations for social policy. But it is exactly this kind of research that one can reasonably predict will be used for racist ends (intentionally and not) in a race-stratified society. (This is an argument not against doing such research but against the refusal to state and discuss publicly the political interests in and possible consequences of the research.)

Does it make sense to refer to this kind of research as objective inquiry when everyone has a stake in its outcome? Moreover, as some social scientists have pointed out, it cannot be value-free to describe such social events as poverty, misery, torture, or cruelty in a value-free way. In the face of those phenomena, every statement counts as either for or against; there is no possible third stance that is value-free. The use of objective language to describe such events results in a kind of pornography; the reader, the observer, consumes for his or her own intellectual satisfaction someone else's pain and misfortune. It is not irrelevant, some critics argue, that scientific method does not appear to provide any criterion for distinguishing whether certain procedures on humans should subsequently be referred to as scientific experiments or as torture.[22]

Defenders of pure science frequently appear to be arguing that a scientist's ignorance of the consequences of his scientific behaviors should be counted as evidence for his objectivity. But if the law finds

21. Gould, *Mismeasure of Man.*
22. See the discussion of this problem in "Commentary by Naomi Scheman" (on Sandra Harding's "The Method Question"), *American Philosophical Association Newsletter on Feminism and Philosophy* 88:3 (1989), 40–44.

avoidable ignorance culpable, why shouldn't science? Of course, no one can guarantee the good consequences of all or perhaps any of one's decisions. But why should it not be regarded as culpable to refuse to consider the consequences of one's acts, as this insistence on the possibility of a separation between pure and applied science directs scientists to do? The "innocence" of science communities—our "innocence"—is extremely dangerous to us all. Perhaps people who have exhibited tendencies toward such innocence should not be permitted to practice science or construct metatheories of science; they are a danger to the already disadvantaged and perhaps even to the species! Why shouldn't we regard ignorance of the reasonably predictable consequences of one's scientific behaviors as evidence not of the objectivity of that research but of incompetence to conduct it? Although I am putting this issue in terms of moral responsibility, it is fundamentally a political issue: how is modern Western science constructed by class, race, and gender struggles? But claiming individual moral responsibility can be a powerful motive for political change.

It is less widely recognized that the technologies science uses in its research processes themselves have political consequences. The use of the telescope moved authority about the heavens from the medieval church to anyone who could look through a telescope. The introduction of complex diagnostic technologies in medical research moves authority about the condition of our bodies from us to medical specialists; in practice, it even tends to move this authority from physicians to lab technicians. These are not trivial involvements of science in political interests and values. Not all technologies can be used in a given society, for the political and social values that a technology expresses or enacts may conflict with the dominant social values. In fact, historians and sociologists of science have pointed out that the technologies of experimental method could not gain widespread acceptability in a slave culture: experimental method requires a trained intellect as well as the willingness to "get one's hands dirty," but slave cultures forbid education to slaves and manual labor to aristocrats.[23]

There is a third important relation between science and technology: scientific problematics are often (some would say always) responses to social needs that have been defined as technological ones. For example, scientists were funded to produce information about the reproductive

23. Zilsel, "Sociological Roots."

system which would permit the development of cheap and efficient contraceptives. The development of contraceptives was a technological solution to what was defined by Western elites as the problem of overpopulation among ethnic and racial minorities in the First World and indigenous Third World peoples. From the perspectives of those peoples lives, however, there are at least equally reasonable ways to define what "the problem" is. Instead of overpopulation, why not talk about the First World appropriation of Third World resources which makes it impossible for the Third World to support its own populations? Why not say that the problem is the lack of education for Third World women—the variable said to be most highly related to high fertility?[24] After all, just one member of a wealthy North American family uses far more of the world's natural resources in his or her daily life than do whole communities of Ethiopians. Would it not be more objective to say that First World overpopulation and greed are primarily responsible for what Westerners choose to call Third World overpopulation?

To take another example, research to develop higher-yield varieties of grains is said to make the Third World better able to feed its peoples. But given the political and economic relations between the First and Third Worlds, what it actually does is to increase the supply of crops for export to the First World, leaving Third World peoples even hungrier than they were before they were the beneficiaries of technological "development." The problem could have been defined as why the First World should profit even further at Third World expense, or who benefits most when the First World so squanders its resources that it needs to import food from far poorer societies.

This argument distinguishes scientists' intentions from the functions of their work. The point is not that scientists *intend* to conduct technology-driven inquiry, or to promote the politics that the production of their information requires or makes possible; most do not. Instead, the point is how scientific research functions within the contemporary social order. This kind of argument is difficult for many people to appreciate because elites—and especially scientists—are taught to think of the results of science as the consequence of individual and

24. See Maria Mies, *Patriarchy and Accumulation on a World Scale: Women in the International Division of Labor* (Atlantic Highlands, N.J.: Zed Books, 1986), for a discussion of why a capitalist imperialist patriarchy paradoxically cannot permit Third World women to reproduce themselves but insists that First World women do so.

team effort to find descriptions of the regularities of nature and their underlying causal tendencies which are less false than the prevailing ones. In such a view, the behaviors of women and members of marginalized races and classes may be regarded as a function of their biological or social characteristics, but not the behavior of elites. Elite behavior is considered the consequence of individual choices and the exercise of will. The contrary argument here depends upon recognition that elite behavior, too, is distinctively shaped by social agendas.

Is there any "pure science" left after we see all these ways in which science and technology are interrelated? Some would say yes—that at least in such projects as the search for the basic constituents of the universe, one can see scientific research that is not technology-driven. Yet this research too uses technologies that themselves have social implications: who is being educated to use them? what kinds of social status accrue to people who get to use these technologies? Moreover, is not apparently pure research often justified on the grounds that it is *likely* to produce technologically useful information? In any case, the cost of producing apparently "useless" information is justifiable to science policymakers on the additional grounds of its halo effect on the rest of science: this 5 percent of "pure research" provides a camouflage for the 95 percent that is so obviously technology-driven. But if that is its function, how is it pure?[25]

Finally, the insistence on the argument for "pure science" may express a deep irrationality about our culture. In a world where so many go hungry, where cities are in decay and countrysides have been devastated, where many need medical assistance they cannot afford, where the literacy gap increases between the haves and the have-nots—where, in short, access to just a few more resources could have such large effects on the lives of so many—in such a world, why should we support scientific activity defined as "pure" precisely *because* it promises no socially usable results? The support of "pure science" might more reasonably be seen as a make-work welfare program for the middle classes in the service of elites. Science is not responsible for all the bad characteristics of contemporary social life, but if it does not

25. See Forman's analysis (in "Behind Quantum Electronics") of loss of purity in twentieth-century physics, and Restivo's argument (in "Modern Science") claiming that the purity of science blocks our ability to understand modern science as a social problem.

develop effective means for identifying the causes and consequences of its own beliefs and practices, it remains complicitous in the production of these social ills. In the insistence that the technologies and applications of the sciences are no part of "science proper," one can locate another false belief that we should give up once and for all. It is no accident that sciences adopting this belief end up disproportionately disadvantaging those, such as women, whom elites define as "other."

(5) "Scientists can provide the most knowledgeable and authoritative explanations of their own activities, so sociologists and philosophers (including feminists) should refrain from making comments about fields in which they are not experts." To many people, it seems obvious that only physicists can really understand the history and practice of physics; only biologists, the reasons why some hypotheses were preferred to others in the history of biology. To hold this view, however, is to hold not the obvious truth that physics should be done by people trained in physics but the quite different belief that the "science of the natural sciences" is best created by natural scientists— of physics by physicists, of chemistry by chemists, and so on. Yet if this were so, the sciences would be the only human activity for which science recommends that the "indigenous peoples" should be given the final word about what constitutes a maximally adequate causal explanation of their lives and works. It would amount to the same thing to say that there cannot be a science of science; that science alone must be exempted from the claim that all human activity and its products— including the content and form of beliefs—can be explained causally. Should we accept this view, then the sciences alone could not be explained in ways that go beyond, or contradict, the understandings its practitioners can produce.

There are at least five reasons why natural scientists are not the best people to provide causal explanations of their own activities (and most of these claims could be adjusted to apply to practitioners in any discipline). In the first place, a science of science will try to locate origins of everyday scientific activity and belief that are not visible from the location of that activity. In some premodern societies, social relations are simple enough to be seen in virtually their entirety from the perspective of everyday life. But in modern societies, social relations are so much more complex that it is impossible to understand how the government, the economy, or the family actually works on the basis of our everyday interactions with and in those institutions.[26] For example, many causes of everyday family life are located far away—in the economy, government policy, Supreme Court decisions, child-rearing practices, religious beliefs, and other aspects of social relations. Similarly, important causes of scientists' everyday activities and experiences are to be found far distant from the laboratory or field site—in the economy, government policy, Supreme Court decisions, child-rearing practices, religious beliefs, and other social relations. A science of science must generate descriptions and explanations of scientific phenomena which start off not in the labs but far away from where scientists and their expertise are located.[27]

In the second place, that "far away" where science begins is temporal as well as spatial. Many patterns in the behaviors of individuals and social institutions are not visible from the single local historical perspective of any individual or any group such as scientists. They are detectable only if one looks systematically over large sweeps of history. At any present moment there appear only confusing and small tendencies in various directions. Patterns in these tendencies appear and accumulate power only over decades or even centuries. Distinctive ways of explaining history will be useful in understanding the causes of everyday life in science. Of course, explaining individual events or processes as parts of larger patterns is one way of describing exactly what natural scientists do. The point is that the history and practices of science themselves can be usefully subjected to such scientific explanations.

But, third, the problem goes still deeper. Scientists' activity as scientists is exactly the wrong kind of activity from which to be able to detect many interesting causal features of science. For one thing, simply by virtue of choosing to continue to carry out the routine practices of this institution, they undermine the probability of their achieving the kind of critical perspective on those practices that "outsiders" could provide (I do not say that they *cannot* provide such a perspective; a few practicing scientists in every field have done so). The same is true of every human activity (including doing philosophy or writing a book). A more important reason, however, is that at least since World War II,

---

26. Dorothy Smith has made this point repeatedly; see *The Everyday World as Problematic: A Feminist Sociology* (Boston: Northeastern University Press, 1987).

27. This problem is neither resolved nor even acknowledged in the work of the "strong programme" theorists; see citations in note 9 above.

doing science has been part of the apparatus of ruling.[28] Science generates capital in the form of information, ideas, and technologies that are used to administer, manage, and control the physical world and social relations. When human activity is divided in hierarchical ways, those who engage in "ruling class" activity can have only a partial and distorted understanding of nature and social relations.[29] For this reason, laboratory life especially is the wrong activity from which to try to describe and explain the causal relations of administering, managing, and controlling the physical world and social relations. Even Kuhn hints at this truth when he points to the false stories about Nobel Prizes and glorious careers in science that scientists generate in order to recruit young people into the arduous training and routine work necessary to careers in science.[30]

In the fourth place, in modern Western cultures, middle-class white men tend more than other groups to believe in the ability of their individual minds to mirror nature, their faculties of judgment to make rational choices, and the power of their wills to bring about their choices. Hence, given the qualities that make them "good scientists," natural scientists are the last people to suppose it desirable to examine the limits of their minds to mirror nature or make rational scientific choices, and of their wills to bring about their choices. They are psychologically the wrong people to provide causal accounts of science. To ask them to try to provide fully causal accounts of their own activity is to ask them to identify the kinds of irrationalities in their own behaviors on which Freud and Marx focused—not to mention the gender and race "irrationalities" identified by later critics.

Finally, natural scientists have the wrong set of professional skills for the project of providing causal accounts of science. What is needed are people trained in critical social theory: that is, in locating the social contexts—psychological, historical, sociological, political, economic —that give meaning and power to historical actors, their ideas, and their audiences. Natural scientists are trained in context-stripping; the

28. See Forman, "Behind Quantum Electronics"; Hilary Rose and Steven Rose, "The Incorporation of Science," in *Ideology of/in the Natural Sciences*, ed. Hilary and Steven Rose (Cambridge, Mass.: Schenkman, 1979).

29. These are the claims of the standpoint theorists, discussed at length in the following chapters.

30. Kuhn, *Structure of Scientific Revolutions*. chap. 11.

science of science, like other social sciences, requires training in context-seeking.

Our ability to understand and explain science would be enhanced if we eliminated the extraneous belief that scientists in general are the best people to describe and explain scientists' activities. This is not to say that they should not be permitted in the group who can provide illuminating accounts of how science works. Scientists, like anyone else, can use causal accounts of science to generate valuable explanations. But they, like anyone else, must learn how to think about and observe sciences and their technologies in ways for which present-day scientific training does not prepare them. They must become critical social scientists to learn how to reflect critically on intuitive, everyday beliefs about methods and nature which further reflection shows are false. For this reason it can be illuminating to think of the natural sciences as inside, part of, social science.

The sciences incorporate both liberatory and oppressive tendencies. They have done so since their origins. The new sciences of the seventeenth century decentered our species from its unique location in a universe described by Christian and feudal thought. They said instead that humans are located on an otherwise ordinary planet circling around an unremarkable sun in an insignificant galaxy and, further, that the earth and the heavens are made up of the same kinds of materials and moved by the same kinds of forces. Thus those new sciences gave antiaristocratic messages. They implied that nature does not specify any essential higher or lower stations in life or human "natures." They undermined belief in the natural legitimacy of royalty and aristocracy. And they were epistemologically antiauthoritarian and participatory. "Anyone can see through my telescope," said Galileo, and can then reason to the conclusions of the new sciences. We are used to thinking in contradictory ways about this particular set of social values carried by modern science. On the one hand, these are thought not to be social values at all, since even though science incorporates them, it can still attain value-neutrality. On the other hand, these values are thought to be so constitutive of science that someone who criticizes science is thought to be against reason, progress, and democracy.

However, the new sciences carried other kinds of social values also. They provided resources for a new social class to assert its legitimacy over others. This class had interests in owning land and developing resources (ores, plants, animals, and the peasants who also belonged to the land) for its own benefit, in using warfare to obtain access to land and resources, and in legitimating only its own activities and achievements as what everyone should recognize as civilization. These interests found a ready companion in the focus of the new sciences on the materiality of the world, on developing more efficient ways to dominate nature, on the value of technological "progress," and on the legitimacy and usefulness of universal laws.

Thus, modern Western science was constructed within and by political agendas that contained both liberatory and oppressive possibilities. Present-day science, too, contains these conflicting impulses. The antidemocratic impulses are not only morally and politically problematic; they also deteriorate the ability of the sciences to provide objective, empirically defensible descriptions and explanations of the regularities and underlying causal tendencies in nature and social relations. One way to focus on this problem is to discover that we have no conception of objectivity that enables us to distinguish the scientifically "best descriptions and explanations" from those that fit most closely (intentionally or not) with the assumptions that elites in the West do not want critically examined. It is only part of the problem that scientists are part of this elite. Without such a strong criterion of objectivity, science can easily become complicitous with the principle that "might makes right," whether or not anyone intends this complicity. The ethics and rationality of science are intimately connected.

(6) "Physics is the best model for the natural sciences, so feminist social science analyses can have nothing to offer the natural sciences." Now we can consider the false belief that produces the title for this chapter. It is still common to regard the natural sciences, and especially physics, as the ideal model for all inquiry. Of course, there is a long history of dispute over whether models of research and explanation originating in the study of inanimate nature are the most useful for studying social beings, but I intend to challenge an assumption made by both sides to that dispute: namely, that the way physics has been taught and practiced—the accepted "logic" of its research processes and forms of explanation—is the best it could be: that "physics" is a good model for physics. Both the "naturalists" and the "intentionalists," as the two parties have been named in the debate over the philosophy of social science, assume that physics provides a perfectly fine model of inquiry and explanation for the natural sciences. That is not

controversial to either group, even though (my point here) it should be. So my argument is not that physics provides a poor model for social inquiry; it is the stronger argument that the paradigm of physics research and explanation, as it is understood by scientists and most other people, is a poor model for physics itself.

We can appreciate the historical reasons why the physics of the seventeenth and subsequent centuries was so highly valued as a model for all scientific inquiry. In the twentieth century the unity-of-science thesis of the Vienna Circle provided the modern justification for prescribing a hierarchy of the sciences with physics at the top. Ironically, my analysis here can be understood to agree that the sciences should be unified—but I propose that the hierarchy should be "stood on its head." On scientific grounds, as well as for moral and political reasons, those social sciences that are most deeply critical and most comprehensively context-seeking can provide the best models for all scientific inquiry, including physics.[31]

It is not helpful from a scientific perspective to take as a model those research projects in which controversy about basic principles is absent—the criterion Thomas Kuhn used to identify research that had reached the truly scientific stage. The problem with Kuhn's criterion is that in sciences that are important to dominant groups in socially stratified societies, lack of controversy about fundamentals is not a reliable or even plausible indicator of the absence of social, economic, and political values. This is such a society, and physics is such a science. Perfect agreement about basic principles and methods of inquiry can be and has often been reached by scientific "guns for hire" employed by the most egregious sexists, imperialists, and profiteers. Even more distressing is the history of well-intentioned research by the most distinguished of scientists which was inadvertently highly constrained by the sexist, racist, imperialist, and bourgeois ethos of its period.[32]

Instead, the model for good science should be research programs explicitly directed by liberatory political goals—by interests in identifying and eliminating from our understanding of nature and social relations the partialities and distortions that have been created by socially coercive projects. It does not *ensure* good empirical results to

31. I discussed this point in a preliminary way in *The Science Question in Feminism*.
32. E.g., the kinds of cases analyzed by Forman, "Behind Quantum Electronics"; and Gould, *Mismeasure of Man*.

select scientific problematics, concepts, hypotheses, and research designs with these goals in mind; democratic sciences must be able to distinguish between how people want the world to be and how it is. But better science is likely to result if *all the causes* of scientific conclusions are thought to be equally reasonable objects of scientific analysis. Since sexism, racism, imperialism, and bourgeois beliefs have been among the most powerful influences on the production of false scientific belief, critical examination of these causes, too, of the "results of research" should be considered to be *inside* the natural sciences. We could say that the natural sciences should be considered to be embedded in the social sciences because everything scientists do or think is part of the social world.

Objections and Responses

The foregoing proposal will seem bizarre to thinkers who are comfortable with the scientific and epistemological authoritarianism embedded in the models of "value-neutral" research that dominate in the natural sciences. Let me respond to some predictable criticisms, even at the risk of repeating in different terms the arguments above.

Here is one: "Who is to decide what is liberatory? What's liberatory for you may not be so for me." It is true that people will have to negotiate through social and political processes about whose lives most deserve improvement at any particular time and, therefore, from the perspective of whose lives sciences should be developed. If those processes are not now sufficiently democratic, then we must take (democratic) steps to make them better. But the problem of "whose perspective?" is not solved by hiding the decision process behind claims of value-neutrality. Many scientists do not really believe—and some actively protest—the dominant scientific ideology. Nevertheless, the myth of experts and their authority is the one used to recruit students into science education and to keep the sciences linked as firmly as possible to the goals of the dominant groups in the West. Thus many people who are most comfortable with hierarchical decision-making and who have little experience in negotiating social arrangements except among white, Western, economically privileged, men like themselves will find it difficult to participate effectively in such negotiations (but it is never too late to learn new skills).

Another objection: "Discussions of the appropriate goals of science should indeed occur, and of course the needs of minorities, women, and the poor should be considered. But there is no good reason to think of these discussions as part of science itself. These are discussions more appropriately conducted in political arenas than in the laboratories and other locations where scientific research is done." Moral and political loyalties, however, have counted as part of the evidence for the best as well as the worst hypotheses in the natural sciences.[33] The problem is not primarily differences between the commitments of individual scientists, for those differences are relatively easy to identify and eliminate from research processes through existing norms of inquiry. The problem, instead, is those values, interests, and commitments that are close to culturewide within scientific cultures or cultural elites, for these cannot even be identified by the methods of the natural sciences. If all the evidence for scientific belief is to be critically examined, so must these social commitments that function as evidence.

Objection: "But I thought it was exactly widespread social beliefs that the individual critical observation and reasoning of the sciences was supposed to correct. It is individuals in the history of the sciences who have formulated hypotheses, observed nature, and interpreted the results of research. The Great Man history of science may not the whole history, but it is a distinguished and central part of it. You are simply proposing that science be entirely subjected to mass thought and thus to the irrationality of politics." But Western scientific thought, no less than the thought of other cultures, has distinctive cultural patterns. I always see through my community's eyes and begin thought with its assumptions. Or, in other words, my society can "observe" the world only through my eyes (and others'), and can begin to think only with my assumptions (and others'). In an important sense, my eyes are not my own, nor are even my most private thoughts entirely private; they belong to my historical period—and to particular class, race, gender, and cultural commitments *that I do not question*. (Questioning, too, belongs to my historical period, but to critical and reflective

---

33. This is another way to put the kind of argument made by Forman, "Behind Quantum Electronics"; Gould, *Mismeasure of Man;* Keller, *Reflections*; Merchant, *Death of Nature;* Van den Daele, "Social Construction of Science," and others.

parts of that history rather than to the "custom and superstition" of the day.) It takes a reorganization of the scientific community and a rethinking of its goals and methods to make visible the social characteristics of the purportedly invisible authors of claims in the natural and social sciences. We need to be able to see how gender, race, and class interests shape laboratory life and the manufacture of scientific knowledge. This, too, is a scientific project, and one that can usefully be regarded as part of the natural sciences.

Objection: "Aren't you arguing that we should substitute subjectivist and relativist stances for objectivity in the sciences?" On the contrary, any research that is conceptualized as maximally value-free on the grounds that—among other things—it does not critically examine the social causes and dimensions of "good" as well as "bad" scientific belief is, I have been arguing, disabled in its attempts to produce objective understandings of nature and social life. It is unable to scrutinize critically one of the significant causes of widespread acceptance of scientific hypotheses without the notion of "strong objectivity" (defined in Chapter 6). Nature causes scientific hypotheses to gain good empirical confirmation, but so, too, does the "fit" of problematics, concepts, and interpretations with prevailing cultural interests and values. A maximally objective understanding of science's location in the contemporary international social order is the goal here. This is far from a call for relativism. Instead it is a call for the maximization of criticism of superstition, custom, and received belief—criticism for which the critical, skeptical attitude of science is supposed to be an important instrument. Ironically, we can have a science of morals and politics not by imitating the natural sciences in designing research in these fields but only by putting critical discussions of morals and politics at the heart of our sciences.

Objection: "Isn't this argument really against science? Aren't you 'down on physics'?" No doubt many will think so. But this argument has a different target. It is against a certain kind of narrow and no longer useful explanation of why it is that physics has contributed so greatly to the growth of scientific knowledge in the West. Only "sciences for the people" (in Galileo's phrase), not for elites, can be justifiably supported in a society committed to democracy. There are plenty of useful projects for such sciences, but they do not include research that provides resources for militarism or for ecological disaster, or continues to move resources away from the underprivileged and toward the already overprivileged.

There is plenty of science still to be done once physics is considered just one human social activity among many others. What kinds of knowledge about the empirical world do we need in order to live at all, and to live more reasonably with one another on this planet from this moment on? Should improving the lives of the few or of the many take priority in answering this question?

II

Epistemology

5

What Is Feminist Epistemology?

Attempts to develop distinctively feminist theories of knowledge have arisen from a number of sources since 1970. As noted in Chapter 2, one important source has been the frustration experienced by social scientists and biologists as they have tried to add women and gender to the existing bodies of knowledge in their fields. The conceptual schemes in these fields and the dominant notions of objectivity, rationality, and scientific method were too weak, or too distorted in some way or another, to be competent even for identifying—let alone eliminating—sexist and androcentric assumptions and beliefs. How could one describe and explain female biology or women's lives within conceptual schemes and research models that so systematically distorted such subject matters? Male biology and men's lives also appeared distorted in significant respects by sexist and androcentric assumptions. Moreover, the feminist perspective from which the identification and criticism of sexism and androcentrism arose was regarded by most scientists as politics (as opposed to reason and observation) and therefore a threat to their purported attempts to provide "pure" descriptions and explanations of the regularities and underlying causal tendencies in nature and social relations.

Since it was the natural sciences that originated the models of objectivity, rationality, and scientific method that are dominant in biology and the social sciences, the suspicion arose that the prevalence of sexism and androcentrism in biology and the social sciences might taint the natural sciences also. In Chapter 4 I pointed to popular but false beliefs about the natural sciences which make it difficult to understand the relevance of feminist criticisms to these fields.

Thus, from a variety of perspectives one could easily come to the conclusion that the concepts of women and of knowledge—socially legitimated knowledge—had been constructed in opposition to each other in modern Western societies. Never had women been given a voice of authority in stating their own condition or anyone else's or in asserting how such conditions should be changed. Never was what counts as general social knowledge generated by asking questions from the perspective of women's lives. In attempting to account for and remedy this situation, several competing feminist epistemologies have been articulated. These theories of knowledge both borrow from and are in tension with "prefeminist" epistemologies, so it requires some careful work to sort out just what feminist epistemologies are claiming and not claiming.

Three main feminist tendencies toward generating new theories of knowledge have emerged: feminist empiricism, feminist standpoint theory, and feminist postmodernism (the last sometimes referred to as anti-Enlightenment, or antihumanist tendencies in feminist thought). This chapter focuses explicitly on the first two, though here and in later chapters I develop the logic of the standpoint theory in ways that more vigorously pull it away from its modernist origins and more clearly enable it to advance some postmodernist goals.[1]

Feminist versus Conventional Epistemology

For many readers, the very idea of a feminist epistemology may appear to be a contradiction in terms. Can the same term—epistemology—really be applied to feminist thought and to the standard way of thinking about epistemology that has dominated U.S. and British philosophy departments in recent decades? Here are excerpts from a representative expression of that Anglo-American view in the 1967 *Encyclopedia of Philosophy:*

Epistemology, or the theory of knowledge, is that branch of philosophy which is concerned with the nature and scope of knowledge, its presup-

1. The semi-rapprochement between standpoint and postmodernist tendencies is a major concern of the rest of the book, esp. Chapters 7, 11, and 12.

positions and basis, and the general reliability of claims to knowledge. . . .

Epistemology differs from psychology in that it is not concerned with why men hold the beliefs that they do or with the ways in which they come to hold them. Psychologists can, in principle, give explanations of why people hold the beliefs they do, but they are not necessarily competent, nor is it their province, to say whether the beliefs are based on good grounds or whether they are sound. The answer to these questions must be sought from those who are experts within the branches of knowledge from which the beliefs are drawn. The mathematician can give the grounds for believing in the validity of Pythagoras' theorem, the physicist can give the grounds for believing in, say, the indeterminacy principle, and an ordinary but reliable witness can provide the grounds for believing in the occurrence of an accident. Normally, when the beliefs are true and the grounds sufficient, it is permissible to claim knowledge, and whether a particular truth can be said to be known may be determined by reference to the grounds which are appropriate to the field from which the truth is drawn. The epistemologist, however, is concerned not with whether or how we can be said to know some particular truth but with whether we are justified in claiming knowledge of some whole class of truths, or, indeed, whether knowledge is possible at all. The questions which he asks are therefore general in a way that questions asked within some one branch of knowledge are not.[2]

This account goes on to point out that the history of Western epistemology has been largely a series of responses to various formulations of the problem of skepticism. The canon reviewed—Plato, Aristotle, Augustine, Aquinas, Ockham, Descartes, Spinoza, Leibniz, Locke, Berkeley, Hume, Kant, Hegel, Bradley, Schopenhauer, Husserl, Mill, Pierce, Dewey, Moore, Russell, logical positivism, Wittgenstein, and "ordinary-language" philosophy—is the canon for the dominant tendency in Western philosophy in general, not just for epistemology. This canon, as noteworthy for what it excludes as for what it includes, is fashioned out of the supposedly fundamental human concern with the problem of whether it is possible for humans to know anything at all.

Can a distinctive feminist epistemology find a home within this particular traditional definition of the field and its problems? At first

2. D. W. Hamlyn, "History of Epistemology," in *Encyclopedia of Philosophy*, vol. 3, ed. Paul Edwards (New York: Macmillan, 1967), 8–9.

glance, one would say no. Feminists who reflect on the social and natural sciences do not appear to be concerned with whether it is possible for humans to know anything at all; consequently, why should there be a distinctively feminist branch of epistemology? In support of this position, one could argue that even though feminist research in biology and the social sciences claims to have discovered some new facts—some new "truths" or, at least, some good candidates for less false beliefs—about females, males, gender, and social relations between the sexes, this research provides just the kind of good old empirical evidence that has been regarded as sufficient for claims to count as plausible, well-supported, less false, better confirmed than their rivals, and so on. In short, one could argue that the results of feminist research can be, and are, justified in familiar and uncontroversial ways. Moreover, although feminists might well explain why sexists or sexist societies hold certain beliefs, these explanations (according to conventional epistemology) would be only psychological or, perhaps, sociological and historical ones. (Anglo-American epistemology has been studiously uninterested in the relevance to philosophy of sociological and historical explanations of belief. It tends to reduce these accounts of "why men hold the beliefs they do or . . . the ways in which they come to hold them" to psychological accounts—as does the *Encyclopedia of Philosophy* entry.) So such a thing as "feminist epistemology"—a specifically feminist theory of knowledge—should be unnecessary. Or has something gone wrong with this line of thought?

One problem is that when evidence that has been found uncontroversial when presented in support of nonfeminist claims is presented to support feminist claims, the uncontroversial becomes controversial. Critics say: "But are these really facts? And are the grounds put forward for their status as facts really reasonable ones? Before rushing prematurely to such a conclusion, shouldn't we at least wait until more rigorous investigation has been conducted, and by more objective observers—or at least by the kinds of researchers conventionally regarded as objective?" Women are inferior at rigorous observation and at reason, such critics are probably thinking, and—worse yet—feminism is a politics. How could women and politics be producing facts that anyone should regard as serious challenges to the impersonal, objective, dispassionate, value-free facts that the natural and social sciences have produced?

Here one can begin to see a place for a distinctively feminist epistemology. It is true that these questioners are not doubting the possibility of any knowledge at all. But neither are they only posing the kind of request for further displays of evidence, grounds for belief, that are characteristic of the usual critical modes *within* biology and social science disciplines. Their questions express not just doubts about some particular belief—"Is there reasonable evidence for claiming that Woman the Gatherer made important contributions to the dawn of distinctively human history?" or "Is the evidence plausible that women's patterns of moral reasoning or of learning are different from men's and equally valuable?"—but general skepticism about the possibility of the results of distinctively feminist research ever being able to achieve the kind of scientific status that the results of nonfeminist research can achieve. Are claims produced by women, or by people whose research is heavily motivated by feminist concerns, really deserving of the term "knowledge" rather than "opinion" or perhaps even "ideology"? Isn't bias the major consequence of taking one's research questions from the lives of a particular social group—especially when that group is women? Since feminism is fundamentally a political movement, aren't its claims biased by its politics? If men disagree with women's claims, why should women's claims ever be considered preferable to men's?

These more general skeptical questions are particular forms of just the kinds of questions conventional epistemology has asked. Who can be subjects, agents, of socially legitimate knowledge? (Only men in the dominant races and classes?) What kinds of tests must beliefs pass in order to be legitimated as knowledge? (Only tests against the dominant group's experiences and observations? Only tests against what men in the ruling groups tend to think of as reliable experience and observation?) What kinds of things can be known? Can "historical truths," socially situated truths, count as knowledge? Should all such situated knowledges be regarded as equally plausible or valid? What is the nature of objectivity? Does it require "point-of-viewlessness"? How can we distinguish between how we want the world to be and how it is if objectivity does not require value-neutrality? What is the appropriate relationship between the researcher and her or his research subjects? Must the researcher be disinterested, dispassionate, and socially invisible to the subject? What should be the purposes of the pursuit of knowledge? Can there be "disinterested knowledge" in a society that is deeply stratified by gender, race, and class? As these questions indicate, the grounds for whole classes of knowledge-claims are at issue in the case of feminist research, not just the "grounds that are appropriate to the field from which the truth is drawn." Thus, feminist research in biology and the social sciences raises distinctively epistemological issues that challenge the conceptual framework—its moral, political, and metaphysical assumptions—within which the dominant Anglo-American epistemology has posed its concerns.

Feminism is not the only intellectual tendency that raises skeptical questions about the still-powerful Anglo-American epistemological tradition. As historians and sociologists of science have pointed out, this tradition would also have difficulty counting as justifiable belief many of the claims that have gained the widest acceptance. Claims about a heliocentric planetary system, the structure of DNA, and the relationship between poverty and the control of labor processes were not legitimated by experts from within fields where the beliefs emerged; there *were* no such fields with their structure of experts. One might even posit that it was precisely observers from outside these fields, as they were then defined, who could bring to their analyses critical perspectives and fresh ways of thinking about the phenomena of interest. As the new histories of science point out, it was exactly the eventually successful efforts to legitimate these claims that created new branches of knowledge and the subsequent development of their "experts." Surely whatever theory of knowledge is regarded as the best, it should be able to explain why the best-supported claims in the history of the sciences did and should have been able to accumulate legitimacy.

The theory of knowledge I have been discussing is not the only epistemological tradition with which feminist theories of knowledge jockey for conceptual space, for legitimacy, but this contemporary Anglo-American epistemology marks important boundaries within which the leading research programs in the natural and social sciences occur. Its conceptual framework and assumptions are very close to the positivism that reflects the "spontaneous consciousness" of science.[3] Chapter 3 briefly described two different approaches to thinking about science and knowledge: feminist empiricism and the feminist standpoint theory. It is time to look at these two epistemologies in more detail.

3. Roy Bhaskar, *Reclaiming Reality* (New York: Verso, 1989), 64.

Feminist Empiricism

Description

Generated by feminist research in biology and the social sciences, feminist empiricism is one epistemological strategy for justifying the challenges to traditional assumptions. In research reports one frequently finds the argument that the sexist and androcentric claims to which the researcher objects are just the result of "bad science." They are caused by social biases, by prejudices. These prejudices are created by hostile attitudes and by false beliefs that are due to superstition, ignorance, or miseducation but have often become entrenched in custom (and sometimes even in law). Such biases enter the research process particularly at the stage when scientific problems are being identified and defined, but they can also appear in the design of research and in the collection and interpretation of data. Feminist empiricists argue that sexist and androcentric biases can be eliminated by stricter adherence to existing methodological norms of scientific inquiry; only "bad science" or "bad sociology" is responsible for their retention in the results of research.

But how can the scientific (biological, sociological, psychological) community come to see that its work *has* been shaped by androcentric prejudices? This is where one can see the importance of movements for social liberation. As Marcia Millman and Rosabeth Moss Kanter have pointed out, the women's movement and others "make it possible for people to see the world in an enlarged perspective because they remove the covers and blinders that obscure knowledge and observation."[4] Furthermore, feminist empiricists often point out, the women's movement creates the opportunity for more women and feminists (male and female) to become researchers, and they are more likely than sexist men to notice androcentric biases.

Though I call this epistemological strategy "feminist empiricism," its practitioners do not usually label it at all; they see themselves as primarily following more rigorously the existing rules and principles of the sciences. They do not see anything particularly noteworthy in their research methods or epistemological approach. Indeed, this talk of

4. Marcia Millman and Rosabeth Moss Kanter, Editors' Introduction to *Another Voice: Feminist Perspectives on Social Life and Social Science* (New York: Anchor Books, 1975), vii.

"bad science" and of more careful data collection did not really become visible as a specifically epistemological move until the feminist standpoint theory began to appear as an alternative to it.

Because I cite criticisms of and problems with feminist empiricism below, for nonphilosophers it needs to be said that these are not criticisms of "doing empirical research" (as some scientists have assumed after reading earlier descriptions of this position). The empiricism at issue in this discussion is a philosophy of science going back at least to Aristotle but associated usually with Locke, Berkeley, Hume, and other British philosophers of the seventeenth and eighteenth centuries. In this sense, empiricism defends experience rather than ideas as the source of knowledge and is thus contrasted with rationalism. Few contemporary scientists or philosophers would want to give such a small role to reason as this definition of empiricism implies—no one is an empiricist of that sort today—yet many natural and social scientists insist on such remnants of empiricist philosophy as the primacy of observations and "pure data" and the necessity of knowing that one has the facts because one's methods are certified by a lineage going straight back to the British empiricists. So while everyone thinks empirical research is important for finding out how nature and social life are organized, one can nevertheless criticize empiricism (and feminist empiricism) as a *theory* about how to do research and to justify its results.

Virtues

The justificatory strategy of feminist empiricism is by no means uncontroversial. Nevertheless, it is often thought to be less threatening to the practices of the sciences and to their epistemologies than the standpoint strategy. It is indeed conservative in several respects, and these constitute in part both its strengths and its weaknesses.

First of all, its appeal is obvious. Many of the claims emerging from feminist research in biology and the social sciences *are* true—or, at least, less false than those they oppose (I recommend that readers have in mind specific claims that have been produced by recent research guided by feminism: claims about "Woman the Gatherer," women's different moral voice, the significance of women's activities in social life, and so on). Within the broad fields of human understanding marked out by the traditional biological and social science disciplines, research intending to reevaluate women's nature and roles and the social dimensions of gender certainly meet overt standards of "good research"—or, at least, as many of these standards as does the (androcentric or sexist) research whose results they contest. The point is not that all feminist claims are automatically preferable because they are feminist but, rather, that when they are found preferable, the preference can be explained in terms of conventional scientific virtues.

Second, feminist empiricism appears to leave intact much of scientists' and philosophers' conventional understanding of the principles of adequate scientific research. It appears to challenge mainly the incomplete practice of the scientific method, not the norms of science themselves. It says, in effect, that mainstream inquiry has not adhered rigorously enough to its own norms. In other words, many scientists can admit that the social values and political agendas of feminism raise issues that enlarge the scope of inquiry and reveal a need for greater care in the conduct of inquiry. They can believe that feminist research leaves the logic of the research process and of scientific explanation still conforming to principles that have always been central to the best "prefeminist" scientific research.

The discourses of objectivity and of truth/falsity are ancient and powerful. It is a great strength of feminist empiricism that it can enter and use these widely respected languages and conceptual schemes. As a justificatory strategy it deserves the label "conservative" in a purely descriptive sense: it conserves, preserves, and saves understandings of scientific inquiry that have been intellectually and politically powerful. It enables the results of feminist research to enter conventional bodies of knowledge and to encounter less resistance in doing so than if less conventional epistemologies were used to justify them. Frequently, however, once they are considered plausible, the feminist claims wreak havoc within networks of traditional belief. One cannot simply "add" feminist claims to those they challenge any more than one can add Copernican to Ptolemaic astronomy: the two sets of beliefs contain tensions and contradictions. But that discovery lies in the future for the conventional researchers and scholars who first countenance the results of feminist research and to whom reports of feminist research in the natural and social sciences have, of necessity, most often been addressed.

Feminist empiricism is relatively persuasive to natural and social scientists and to conventional philosophers of science, and—all other things being equal (such as the distance of one's claims from falsity)—a persuasive argument is definitely the best kind to have! The point of a theory of knowledge is not that it be correct by some unfamiliar standard that most people don't accept but that it be persuasive to reasonable, thoughtful, and informed listeners. When conventional natural scientists, social scientists, and philosophers of science are the audiences, feminist empiricism frequently is the justificatory strategy that best meets this criterion.

Such audiences are not the only ones to which feminist researchers present the results of their research. Some others—such as Marxists, and certain groups within the sociology of knowledge and political theory—are already skeptical of the way scientists understand their own activities and thus do not find the empiricist arguments entirely convincing. But other justificatory strategies can be used for these audiences. Why should we limit our strategies to only one when plausibility, not a mystically transhistorical epistemology, is the goal? In this respect, epistemologies are like the models, analogies, and metaphors so important to scientific explanations themselves. When the goal is to increase our understanding of nature, there is no point in using a metaphor that few people understand. "Nature is a machine" became a fruitful metaphor only when people began to gain familiarity with mechanical devices. Similarly, if one wants to explain why feminists can provide fresh perspectives on familiar materials, or why the women's movement is a valuable scientific resource (in addition to its moral and political virtues), there is little point in invoking a controversial strategy for people who don't understand it. I am not recommending a cynical attitude to this feminist empiricist theory of knowledge, even though, as I shall shortly argue, it is not the most satisfactory one to use in understanding feminist research. I am proposing, instead, that many of its central assumptions and claims are not false and that it is effective at explaining the successes of feminist-inspired research to certain important audiences for this work.

Third, the conservatism of feminist empiricism has a related advantage. The social structure of the sciences remains hostile to women scientists, especially to researchers engaged in learning more about women and gender in ways directed by antisexist assumptions. A conservative justificatory strategy is the most robust one that many such scientists can manage and still maintain the disciplinary respect necessary for their continuing access to funding and to teaching and laboratory appointments. As the historians of science remind us, it has taken more than a century of struggle to produce for women these "privileges" that are so much more readily available to men.[5] It certainly would be arrogant of philosophers and other science observers who are not involved in the day-to-day survival battles in the laboratories and social science projects to judge as inadequate, inappropriate, too deeply flawed, or naive those justificatory strategies that feminist researchers have successfully used in order to continue their research.

A fourth conservative virtue of feminist empiricism is that it can be strengthened by appeal to "the ancients"—frequently a useful strategy in the face of disbelief. One can appeal to the forces responsible for the origins of modern science itself as well as to widely recognized moments of later scientific growth to increase the plausibility of this kind of claim. After all, wasn't it the bourgeois revolution of the fifteenth to seventeenth centuries that permitted early modern thinkers to see the world in an enlarged perspective? Many argue that the great social revolution from feudalism to modernism was crucial in helping to remove the covers and blinders that obscured earlier knowledge-seeking and observation. Furthermore, the proletarian revolutions of the late nineteenth century were responsible for yet one more leap in the objectivity of knowledge claims, permitting for the first time an understanding of the effects of class struggles on social relations. Finally, one could point to the twentieth-century deconstruction of European and North American colonialism and its obvious positive effects on the growth of scientific knowledge. As biologist and historian of science Stephen Jay Gould has put the point: "Science, since people must do it, is a socially embedded activity. . . . Much of its change through time does not record a closer approach to absolute truth, but the alteration of cultural contexts that influence it so strongly."[6] From these historical perspectives, the contemporary women's movement, occurring on an international scale, is just the most recent of these revolutions, each of which moves us yet closer to the goals of the creators of modern science.

Beyond the Paternal Discourse

Further consideration of feminist empiricism reveals that despite its conservatism, its feminist component deeply undercuts the assump-

---

5. See Margaret Rossiter, *Women Scientists in America: Struggles and Strategies to 1940* (Baltimore, Md.: Johns Hopkins University Press, 1982).
6. Stephen Jay Gould, *The Mismeasure of Man* (New York: Norton, 1981), 21–22.

tions of traditional empiricism in three ways: feminist empiricism has a radical future.[7] In the first place, it seems to suggest strongly that the "context of discovery" is just as important as the "context of justification" for eliminating social biases that contribute to partial and distorted explanations and understandings. The social context of "before" or "after" a feminist movement makes a difference to the "goodness" of the results of research. After a women's movement, everyone can see things that could not be seen before (and some people can see them more clearly than others). Thus, the individualism of empiricism and its paternal liberal political theory is challenged by feminist epistemology. What we can see in the world around us is a function not just of what is there plus our individual talents and skills but of how our society designs the cultural filters through which we observe the world around and within us and how it institutionalizes those filters in ways that leave them invisible to individuals. Individual biases and differences in assumptions can be identified and eliminated by routine scientific methods, but the culturewide ones require different methods of detection. A social movement on behalf of the less advantaged groups is one such different and valuable "scientific method."

Second, feminist empiricism makes the related claim that scientific method is insufficient to eliminate some kinds of social biases, such as androcentrism, especially when androcentrism arrives in the inquiry process through the identification and definition of research problems. Conventional empiricism holds that scientific method will eliminate any social biases as a hypothesis goes through its rigorous tests. But feminist empiricism argues that an androcentric picture of nature and social life emerges from sciences that do not take feminist concerns seriously. (This issue, like the preceding one, tends to be put in terms of differences between women and men researchers: androcentric explanations emerge from the testing, by men only, of hypotheses generated by what men find problematic in the world around them. But the issue is clearer if it is put in terms of the relationship of feminist politics to scientific research.) The problem is not only that those hypotheses which would most deeply challenge androcentric beliefs are missing from the set of alternative hypotheses available to be considered in the

---

7. As Zillah Eisenstein pointed out about the liberal feminist political theory with which feminist empiricism is linked; see *The Radical Future of Liberal Feminism* (New York: Longman, 1981).

absence of feminist thought. It is also that traditional empiricism does not direct researchers to locate their scientific projects in the same critical plane as their subject matters (a topic I shall pursue further). Consequently, when nonfeminist research gathers evidence for or against hypotheses, "scientific method"—lacking such a directive—is impotent to locate and eradicate the androcentrism that shapes the research process and the picture of nature and social relations that emerges from it.

Finally, although on the one hand feminist empiricists often exhort social scientists to follow existing research methods and norms more rigorously, on the other hand they can also be understood as arguing that following these norms is precisely what contributes to androcentric research results. The norms themselves have been constructed primarily to produce answers to the kinds of questions an androcentric society has about nature and social life, and to prevent scrutiny of the way beliefs that are nearly or completely culturewide in fact cannot be eliminated from the results of research by these norms. A reliable picture of women's worlds and of social relations between the sexes often requires alternative approaches to inquiry that challenge traditional research habits. It is not only that the underlying general principles of scientific method are not powerful enough to detect culturewide sexist and androcentric biases but also that the particular methods and norms of the special sciences are themselves sexist and androcentric.

Thus feminist empiricism intensifies recent tendencies in the philosophy and social studies of science to problematize empiricist epistemological assumptions.[8] There is a tension between empiricist epistemology and its uses by feminists. Some biologists and social scientists, however, have used other resources to justify the new research on women and gender, thereby creating feminist standpoint epistemology.

---

8. See, e.g., Bhaskar, *Reclaiming Reality;* Steven Fuller, *Social Epistemology* (Bloomington: Indiana University Press, 1988); Joseph Rouse, *Knowledge and Power: Toward a Political Philosophy of Science* (Ithaca: Cornell University Press, 1987). Lynn Hankinson Nelson manages to problematize familiar empiricist epistemological assumptions and yet simultaneously to make feminist empiricism a more defensible position by arguing that the individualism many empiricists have assumed is really inconsistent with empiricism. She argues that empiricism needs feminism and its critique of individualism to succeed at its own projects. See her *Who Knows: From Quine to a Feminist Empiricism* (Philadelphia: Temple University Press, 1990).

Before turning to this theory, I must note a transitional position. More radical than the feminist criticisms of "bad science" are those that take as their target Western generalizations from masculine to human in the case of ideal reason. For standpoint theorists (discussed below), such a critique provides one motivation for the development of a specifically feminist theory of knowledge. Other thinkers who participate in this critique stop short of standpoint theory, either because they specifically do not intend to develop such a program or because they have other things on their minds.[9] Philosophers such as Genevieve Lloyd, Sara Ruddick, and Susan Bordo and historians of science such as Evelyn Fox Keller criticize what has come to be called "abstract masculinity."[10] They point out that ideals of Western rationality, including scientific thought, distort and leave partial our understandings of nature and social relations by devaluing contextual modes of thought and emotional components of reason.

Empirical support for their criticism is provided by psychologists. Best known is Carol Gilligan's study of women's moral reasoning.[11] Since scientific reason includes normative judgments (which is the most interesting or potentially fruitful hypothesis or research program to pursue?), the import of Gilligan's work for philosophy is not restricted to ethics. More recently, Mary Belenky and her colleagues, in investigating developmental patterns in women's thinking about reason and knowledge, have pointed to gender bias in philosophic and scientific ideals and suggested its origins in gendered experience.[12] These critical tendencies provide additional grounds for questioning empiricist assumptions.

9. Some people use the terms "standpoint" and "perspective" interchangeably, so that they refer to themselves and others as holding standpoint theories when in fact the theories they speak of are empiricist, or pluralist, or otherwise different from the epistemology developed by "standpoint theorists."

10. See Genevieve Lloyd, *The Man of Reason: "Male" and "Female" in Western Philosophy* (Minneapolis: University of Minnesota Press, 1984); Sara Ruddick, "Maternal Thinking," *Feminist Studies* 6:2 (1980); Sara Ruddick, *Maternal Thinking: Toward a Politics of Peace* (Boston: Beacon Press, 1989); Susan Bordo, *The Flight to Objectivity* (Albany: State University of New York Press, 1987); Evelyn Fox Keller, *Reflections on Gender and Science* (New Haven, Conn.: Yale University Press, 1984).

11. Carol Gilligan, *In a Different Voice: Psychological Theory and Women's Development* (Cambridge, Mass.: Harvard University Press, 1982).

12. Mary G. Belenky, B. M. Clinchy, N. R. Goldberger, and J. M. Tarule, *Women's Ways of Knowing: The Development of Self, Voice, and Mind* (New York: Basic Books, 1986).

Feminist Standpoint Epistemology

Description

A second response to the question about how to justify the results of feminist research is provided by the feminist standpoint theorists. They argue that not just opinions but also a culture's best beliefs—what it calls knowledge—are socially situated. The distinctive features of women's situation in a gender-stratified society are being used as resources in the new feminist research. It is these distinctive resources, which are not used by conventional researchers, that enable feminism to produce empirically more accurate descriptions and theoretically richer explanations than does conventional research. Thus, the standpoint theorists offer an explanation different from that of feminist empiricists of how research directed by social values and political agendas can nevertheless produce empirically and theoretically preferable results.

Just who are these "standpoint theorists"? Three in particular have made important contributions: Dorothy Smith, Nancy Hartsock, and Hilary Rose.[13] In addition, Jane Flax's early work developed standpoint themes; Alison Jaggar used standpoint arguments in her *Feminist Politics and Human Nature,* and I developed briefly one version of this theory and later discussed the emergence of a number of them in *The Science Question in Feminism.*[14] Standpoint arguments are also implicit and, increasingly, explicit in the work of many other feminist thinkers.[15]

13. See Dorothy Smith, *The Everyday World as Problematic: A Feminist Sociology* (Boston: Northeastern University Press, 1987), a collection of the essays that Smith began to publish in the mid-1970s; Nancy Hartsock, "The Feminist Standpoint: Developing the Ground for a Specifically Feminist Historical Materialism," in *Discovering Reality,* ed. Sandra Harding and Merrill Hintikka (Dordrecht: Reidel, 1983); Hilary Rose, "Hand, Brain, and Heart: A Feminist Epistemology for the Natural Sciences," *Signs* 9:1 (1983).

14. Jane Flax, "Political Philosophy and the Patriarchal Unconscious: A Psychoanalytic Perspective on Epistemology and Metaphysics," in Harding and Hintikka, *Discovering Reality;* Alison Jaggar, *Feminist Politics and Human Nature* (Totowa, N.J.: Rowman & Allenheld, 1983), esp. chap. 11; Sandra Harding, "Why Has the Sex-Gender System Become Visible Only Now?" in Harding and Hintikka, *Discovering Reality;* Sandra Harding, *The Science Question in Feminism* (Ithaca: Cornell University Press, 1986), chap. 6.

15. Standpoint theory arguments have been made in the context of other liberatory social movements as well (a point to which I return later): see, e.g., Samir Amin,

This justificatory approach originates in Hegel's insight into the relationship between the master and the slave and the development of Hegel's perceptions into the "proletarian standpoint" by Marx, Engels, and Georg Lukács. The assertion is that human activity, or "material life," not only structures but sets limits on human understanding: what we do shapes and constrains what we can know. As Hartsock argues, if human activity is structured in fundamentally opposing ways for two different groups (such as men and women), "one can expect that the vision of each will represent an inversion of the other, and in systems of domination the vision available to the rulers will be both partial and perverse."[16]

The feminist standpoint theories focus on gender differences, on differences between women's and men's situations which give a scientific advantage to those who can make use of the differences. But what are these differences? On what grounds should we believe that conventional research captures only "the vision available to the rulers"? Even if one is willing to admit that any particular collection of research results provides only a partial vision of nature and social relations, isn't it going too far to say that it is also perverse or distorted?[17] What is it

---

Eurocentrism (New York: Monthly Review Press, 1989); Bettina Aptheker, *Tapestries of Life: Women's Work, Women's Consciousness, and the Meaning of Daily Life* (Amherst: University of Massachusetts Press, 1989); Patricia Hill Collins, "Learning from the Outsider Within: The Sociological Significance of Black Feminist Thought," *Social Problems* 33 (1986); Walter Rodney, *How Europe Underdeveloped Africa* (Washington, D.C.: Howard University Press, 1982); Edward Said, *Orientalism* (New York: Pantheon Books, 1978); Edward Said, Foreword to *Selected Subaltern Studies*, ed. Ranajit Guha and Gayatri Chakravorty Spivak (New York: Oxford University Press, 1988), viii.

16. Hartsock, "The Feminist Standpoint," p. 285. I have written about the distinction between feminist empiricist and feminist standpoint epistemologies reported above in a number of places, usually in order to set the scene for a discussion of the less familiar standpoint epistemology. The most developed of such accounts are "Feminism and Theories of Scientific Knowledge," *American Philosophical Association Feminism and Philosophy Newsletter* 1 (1987); "Epistemological Questions," Editor's Conclusion to *Feminism and Methodology: Social Science Issues*, ed. Sandra Harding (Bloomington: Indiana University Press, 1987); and "Feminist Justificatory Strategies," in *Women, Knowledge, and Reality: Explorations in Feminist Philosophy*, ed. Ann Garry and Marilyn Pearsall (Boston: Unwin Hyman, 1989). The following analysis of why gender differences create scientific and epistemological resources is a slightly revised version of "Starting Thought from Women's Lives: Eight Resources for Maximizing Objectivity," *Journal of Social Philosophy* 21 (1990).

17. I have substituted "distorted" for "perverse," since one person's "perversities" may be another's most highly valued pleasures. "Distorted" appears less amenable to this kind of transvaluation.

about the social situation of conventional researchers that is thought to make their vision partial and distorted? Why is the standpoint of women—or of feminism—less partial and distorted than the picture of nature and social relations that emerges from conventional research?

We can identify many differences in the situations of men and women that have been claimed to provide valuable resources for feminist research. These can be thought of as the "grounds" for the feminist claims.[18]

(1) Women's different lives have been erroneously devalued and neglected as starting points for scientific research and as the generators of evidence for or against knowledge claims. Knowledge of the empirical world is supposed to be grounded in that world (in complex ways). Human lives are part of the empirical world that scientists study. But human lives are not homogeneous in any gender-stratified society. Women and men are assigned different kinds of activities in such societies; consequently, they lead lives that have significantly different contours and patterns. Using women's lives as grounds to criticize the dominant knowledge claims, which have been based primarily in the lives of men in the dominant races, classes, and cultures, can decrease the partialities and distortions in the picture of nature and social life provided by the natural and social sciences.[19]

Sometimes this argument is put in terms of personality structures. Jane Flax and other writers who draw on object relations theory point to the less defensive structure of femininity than of masculinity. Different infantile experiences, reinforced throughout life, lead men to perceive their masculinity as a fragile phenomenon that they must continually struggle to defend and maintain. In contrast, women perceive femininity as a much sturdier part of the "self." Stereotypically, "real women" appear as if provided by nature; "real men" appear as a fragile social construct. Of course, "typical" feminine and masculine personality structures are different in different classes, races, and cultures. But insofar as they are different from each other, it deteriorates objectivity to devalue or ignore what can be learned by starting

18. In *The Science Question in Feminism,* I discussed differences between the grounds proposed by four standpoint theorists: Rose, Hartsock, Flax, and Smith. Here I consider additional grounds proposed to justify feminist research.

19. Standpoint theories need not commit essentialism. *The Science Question in Feminism* contributed to such a misreading of their "logic"; in this book I contest an essentialist reading.

research from the perspective provided by women's personality structures.[20]

Sometimes this argument is put in terms of the different modes of reasoning that are developed to deal with distinctive kinds of human activity. Sara Ruddick draws our attention to the "maternal thinking" that is characteristic of people (male or female) who have primary responsibility for the care of small children. Carol Gilligan identifies those forms of moral reasoning typically found in women's thought but not found in the dominant Western "rights orientation" of ethics. And Mary Belenky and her colleagues argue that women's ways of knowing exhibit more generally the concern for context that Gilligan sees in moral knowing.[21]

One could argue also that the particular forms of any emotion that women experience as an oppressed, exploited, and dominated gender have a distinctive content that is missing from all those parallel forms in their brothers' emotional life. Consider suffering, for example. A woman suffers not only as a parent of a dying child, as a child of sick parents, as a poor person, or as a victim of racism. Women suffer in ways peculiar to *mothers* of dying children, to *daughters* of sick parents, to poor *women*, and in the special ways that racist policies and practices affect *women's* lives. Mother, daughter, poor woman, and racially oppressed woman are "nodes" of historically specific social practices and social meanings that mediate when and how suffering occurs for such socially constructed persons. Women's pleasures, angers, and other emotions too are in part distinctive to their social activities and identities as historically determinate women, and these provide a missing portion of the human lives that human knowledge is supposed to be both grounded in and about.

Whatever the kind of difference identified, the point of these arguments is that women's "difference" is only difference, not a sign of inferiority. The goal of maximizing the objectivity of research should

20. Flax, "Political Philosophy." See also Nancy Hirschmann's use of object relations theory to ground a standpoint epistemology in her "Freedom, Recognition, and Obligation: A Feminist Approach to Political Theory," *American Political Science Review* 83:4 (1989).

21. Ruddick, *Maternal Thinking;* Gilligan, *In a Different Voice;* Belenky et al., *Women's Ways of Knowing.* I said above that Ruddick, Gilligan, and Belenky—among others—do not develop their criticisms of the generalization from stereotypically masculine to paradigmatically human reason into a standpoint epistemology. My point here is that their arguments can be used to do so.

require overcoming excessive reliance on distinctively masculine lives and making use also of women's lives as origins for scientific problematics, sources of scientific evidence, and checks against the validity of knowledge claims.

Some thinkers have assumed that standpoint theories and other kinds of justifications of feminist knowledge claims must be grounded in women's *experiences*. The terms "women's standpoint" and "women's perspective" are often used interchangeably, and "women's perspective" suggests the actual perspective of actual women—what they can in fact see. But it cannot be that women's experiences in themselves or the things women say provide reliable grounds for knowledge claims about nature and social relations. After all, experience itself is shaped by social relations: for example, women have had to *learn* to define as rape those sexual assaults that occur within marriage. Women had experienced these assaults not as something that could be called rape but only as part of the range of heterosexual sex that wives should expect.

Moreover, women (feminists included) say all kinds of things—misogynist remarks and illogical arguments; misleading statements about an only partially understood situation; racist, class-biased, and heterosexist claims—that are scientifically inadequate. (Women, and feminists, are not worse in this respect than anyone else; we too are humans.) Furthermore, there are many feminisms, and these can be understood to have started their analyses from the lives of different historical groups of women: liberal feminism from the lives of women in eighteenth- and nineteenth-century European and American educated classes; Marxist feminism from the lives of working-class women in nineteenth- and twentieth-century industrializing societies; Third World feminism from late twentieth-century Third World women's lives. Moreover, we all change our minds about all kinds of issues. So while both "women's experiences" and "what women say" certainly are good places to begin generating research projects in biology and social science, they would not seem to be reliable grounds for deciding just which claims to knowledge are preferable.

For a position to count as a standpoint, rather than as a claim—equally valuable but for different reasons—for the importance of listening to women tell us about their lives and experiences, we must insist on an objective location—women's lives—as the place from which feminist research should begin. We would not know to value that location so highly if women had not insisted on the importance of their experiences and voices. (Each woman can say, "I would not know to value my own experience and voice or those of other women if women had not so insisted on the value of women's experiences and voices.") But it is not the experiences or the speech that provides the grounds for feminist claims; it is rather the subsequently articulated observations of and theory about the rest of nature and social relations—observations and theory that start out from, that look at the world from the perspective of, women's lives. And who is to do this "starting out"? With this question it becomes clear that knowledge-seeking requires democratic, participatory politics. Otherwise, only the gender, race, sexuality, and class elites who now predominate in institutions of knowledge-seeking will have the chance to decide how to start asking their research questions, and we are entitled to suspicion about the historic location from which those questions will in fact be asked. It is important both to value women's experiences and speech and also to be able to specify carefully their exact role in the production of feminist knowledges.

(2) Women are valuable "strangers" to the social order. Another basis claimed for feminist research by standpoint thinkers is women's exclusion from the design and direction of both the social order and the production of knowledge. This claim is supported by the sociological and anthropological notion of the stranger or outsider. Sociologist Patricia Hill Collins summarizes the advantages of outsider status as identified by sociological theorists. The stranger brings to her research just the combination of nearness and remoteness, concern and indifference, that are central to maximizing objectivity. Moreover, the "natives" tend to tell a stranger some kinds of things they would never tell each other; further, the stranger can see patterns of belief or behavior that are hard for those immersed in the culture to detect.[22] Women are just such outsiders to the dominant institutions in our society, including the natural and social sciences. Men in the dominant groups are the "natives" whose life patterns and ways of thinking fit all too closely the dominant institutions and conceptual schemes.

In the positivist tendencies in the philosophy of the social sciences, these differences between the stranger and the natives are said to measure their relative abilities to provide causal explanations of the natives'

22. Collins, "Learning from the Outsider Within," S15.

beliefs and behaviors. Only understanding, not explanation, can result from the natives' own accounts of their beliefs and behaviors, or from the accounts of anthropologists or sociologists who "go native" and identify too closely with the natives. Because women are treated as strangers, as aliens—some more so than others—by the dominant social institutions and conceptual schemes, their exclusion alone provides an edge, an advantage, for the generation of causal explanations of our social order from the perspective of their lives. Additionally, however, feminism teaches women (and men) how to see the social order from the perspective of an outsider. Women have been told to adjust to the expectations of them provided by the dominant institutions and conceptual schemes. Feminism teaches women (and men) to see male supremacy and the dominant forms of gender expectations and social relations as the bizarre beliefs and practices of a social order that is "other" to us. *It* is "crazy"; we are not.

This claim about the grounds for feminist research also captures the observation of so many sociologists and psychologists that the social order is dysfunctional for women. There is a closer fit for men in the dominant groups between their life needs and desires and the arrangement of the social order than there is for any women. But this kind of claim has to be carefully stated to reflect the extremely dysfunctional character of the U.S. social order for men who are *not* members of dominant groups—for example, African Americans and Hispanics. It is clearly more dysfunctional for unemployed African American and Hispanic men than it is for economically privileged white women. Nevertheless, with extremely important exceptions, this insight illuminates the comparison of the situation of women and men in many of the same classes, races, and cultures. It also captures the observation that within the same culture there is in general a greater gap for women than for men between what they say or how they behave, on the one hand, and what they think, on the other hand. Women feel obliged to speak and act in ways that inaccurately reflect what they would say and do if they did not so constantly meet with negative cultural sanctions. The socially induced need for women always to consider "what men (or 'others') will think" leads to a larger gap between their observable behavior and speech and their thoughts and judgments.

(3) Women's oppression gives them fewer interests in ignorance. The claim has been made that women's oppression, exploitation, and domination are grounds for transvaluing women's differences because

members of oppressed groups have fewer interests in ignorance about the social order and fewer reasons to invest in maintaining or justifying the status quo than do dominant groups. They have less to lose by distancing themselves from the social order; thus, the perspective from their lives can more easily generate fresh and critical analyses. (Women have less to lose, but not nothing to lose; gaining a feminist consciousness is a painful process for many women.)

This argument can be put in terms of what women, and especially feminist women, can come to be willing to say. But it is less confusing if it is put in terms of what can be seen if we start thinking and researching from the perspective of the lives of oppressed people. The understanding that they are oppressed, exploited, and dominated—not just made miserable by inevitable natural or social causes—reveals aspects of the social order that are difficult to see from the perspective of their oppressors' lives. For example, the perception that women believe they are firmly saying no to certain sexual situations in which men consistently perceive them to have said yes or "asked for it" (rape, battering) becomes explainable if one believes that there can never be objectively consensual relations between members of oppressor and oppressed groups. It is from the perspective of women's interests that certain situations can be seen as rape or battering which from the perspective of the interests of men and the dominant institutions were claimed to be simply normal and desirable social relations between the sexes.

(4) Women's perspective is from the other side of the "battle of the sexes" that women and men engage in on a daily basis. "The winner tells the tale," as historians point out, and so trying to construct the story from the perspective of the lives of those who resist oppression generates less partial and distorted accounts of nature and social relations.

Far from being inert "tablets"—blank or not—human knowers are active agents in their learning. Knowledge emerges for the oppressed through the struggles they wage against their oppressors. It is because women have struggled against male supremacy that research starting from their lives can be made to yield up clearer and more nearly complete visions of social reality than are available only from the perspective of men's side of these struggles. "His resistance is the measure of your oppression" said the early 1970s slogan that attempted to explain why it was that men resisted so strenuously the housework, child care, and other "women's work" that they insisted was so easy and required so few talents and so little knowledge.

As I put the point earlier, knowledge is produced through "craft" procedures, much as a sculptor comes to understand the real nature of the block of marble only as she begins to work on it. The strengths and weaknesses of the marble—its unsuspected cracks or surprising interior quality—are not visible until the sculptor tries to give it a shape she has in mind. Similarly, we can come to understand hidden aspects of social relations between the genders and the institutions that support these relations only through struggles to change them. Consider an example from the history of science: it is only because of the fierce struggles waged in the nineteenth and early twentieth centuries to gain formal equality for women in the world of science that we can come to understand that formal equality is not enough. As Margaret Rossiter points out, all the formal barriers to women's equity in education, credentialing, lab appointments, research grants, and teaching positions have been eliminated, yet there are still relatively few women to be found as directors and designers of research enterprises in the natural sciences.[23] The struggles to end discrimination against women in the sciences enabled people to see that formal discrimination was only the front line of defense against women's equity in scientific fields.

Hence, feminist politics is not just a tolerable companion of feminist research but a necessary condition for generating less partial and perverse descriptions and explanations. In a socially stratified society the objectivity of the results of research is increased by political activism by and on behalf of oppressed, exploited, and dominated groups. Only through such struggles can we begin to see beneath the appearances created by an unjust social order to the reality of how this social order is in fact constructed and maintained. This need for struggle emphasizes the fact that a feminist standpoint is not something that anyone can have simply by claiming it. It is an achievement. A standpoint differs in this respect from a perspective, which anyone can have simply by "opening one's eyes." Of course, not all men take the "men's position" in these struggles; there have always been men who joined women in working to improve women's conditions, just as there have always been women who—whatever their struggles with men in their private lives—have not thought it in their interest to join the collective

23. Rossiter, *Women Scientists in America.*

and institutional struggles against male supremacy. Some men have been feminists, and some women have not.

(5) Women's perspective is from everyday life. A fifth basis for the superiority of starting research from the lives of women rather of men in the dominant groups has been pointed out in one form or another since the early 1970s. The perspective from women's everyday activity is scientifically preferable to the perspective available only from the "ruling" activities of men in the dominant groups. Dorothy Smith has developed this argument most comprehensively: women have been assigned the kinds of work that men in the ruling groups do not want to do, and "women's work" relieves these men of the need to take care of their bodies or of the local places where they exist, freeing them to immerse themselves in the world of abstract concepts. The labor of women "articulates" and shapes these men's concepts of the world into those appropriate for administrative work.[24] Moreover, the more successfully women perform "women's work," the more invisible it becomes to men. Men who are relieved of the need to maintain their own bodies and the local places where they exist come to see as real only what corresponds to their abstracted mental world. This is why men see "women's work" not as real human activity—self-chosen and consciously willed (even within the constraints of a male-dominated social order)—but only as natural activity, a kind of instinctual labor such as bees and ants perform. Women are thus excluded from men's conceptions of culture and history.

For Smith, a sociologist, the discipline of sociology has played a major role in "shaping up" the events and processes of everyday life into the kinds of phenomena that can be processed by the administrators and managers who constitute the "rulers" in contemporary Western societies. The dominant conceptual schemes in sociology "articulate" everyday experiences to administrative and managerial agendas and practices. Starting from the "standpoint of women" in this organization of social activity enables us to recover the processes through which social life in fact has taken the forms we see around us.[25] Smith focuses on her own discipline, but sociology is not the only area of research that assists in making the fit between the activities of

24. Smith, *Everyday World.* See Hartsock's similar argument in "The Feminist Standpoint."
25. Smith writes of a "standpoint of women," not a feminist standpoint.

men in the ruling groups and the dominant conceptual schemes used to administer and manage society.

Historian Bettina Aptheker, drawing on the analyses of Smith and others, shows that if we start from the "dailiness" women's lives, we will come to some understandings of both women's and men's lives very different from the accounts favored in conventional social theory.

By the dailiness of women's lives I mean the patterns women create and the meanings women invent each day and over time as a result of their labors and in the context of their subordinated status to men. The point is not to describe every aspect of daily life or to represent a schedule of priorities in which some activities are more important or accorded more status than others. The point is to suggest a way of knowing from the meanings women give to their labors. The search for dailiness is a method of work that allows us to take the patterns women create and the meanings women invent and learn from them. If we map what we learn, connecting one meaning or invention to another, we begin to lay out a different way of seeing reality. This way of seeing is what I refer to as women's standpoint.[26]

Consider what we can learn about resistance to oppression and domination from Aptheker's account of the search for dailiness. She points out that women have often been blamed for collaboration with the dominant groups and for accommodation to oppressive conditions. As a conservative force in history, women should not be relied upon to see it as in their own interests to resist domination and to engage in revolutionary politics—or so this story goes. However, notes Aptheker, men have reserved for themselves the right to define what counts as significant resistance and what counts as collaboration or accommodation. They have done so from the perspective of the kinds of public and political activities from which women have until recently largely been excluded. Moreover, feminists have tended toward the same assessments of women's history of resistance. As Aptheker observes, many of us have come to our feminisms from struggles to participate as equals within conventionally defined political movements. Consequently, we are little different from "malestream" historians and the men in these movements in that we have failed to appreciate a kind of women's resistance that is not "feminist," "socialist,"

26. Aptheker, *Tapestries of Life*, 39.

"radical," or "liberal" but is nevertheless central to the making of history and a source of social change.

Aptheker discusses many ways in which women's struggles to improve the quality of daily life and preserve a cultural heritage constitute important strategies of political resistance to oppression and domination. In labor struggles, the family and community networks that women have forged in their daily lives have sometimes proved more important than the unions in securing better conditions for workers. Women's resistance on behalf of their children—to poverty, to social agencies of the dominant culture, to slavery and concentration camps, to molesting and abusive husbands and fathers—have made survival possible for people who, in poet Audre Lorde's phrase, were "never meant to survive."[27] From the perspective of the dailiness of women's lives, conventional assumptions—of the opposition between social and private injustices, of the opposition between resistance and collaboration, and of social change as occurring only through movements for political power—can be seen to block our ability to understand both women's lives and history.

(6) Women's perspective comes from mediating ideological dualisms: natures versus culture. Other standpoint theorists have stressed the ways in which women's activities mediate the divisions and separations in contemporary Western cultures between nature and culture and such manifestations of this polarity as intellectual work, on the one hand, and manual or emotional work, on the other hand. For example, as Nancy Hartsock has noted,

> women's labor, like that of the male worker, is contact with material necessity. Their contribution to subsistence, like that of the male worker, involves them in a world in which the relation to nature and to concrete human requirements is central, both in the form of interaction with natural substances whose quality, rather than quantity, is important to the production of meals, clothing, etc., and in the form of close attention to the natural changes in these substances. Women's labor both for wages and even more in household production involves a unification of mind and body for the purpose of transforming natural substances into socially defined goods. This too is true of the labor of the male worker.[28]

27. Audre Lorde, "A Litany for Survival," in *The Black Unicorn* (New York: Norton, 1978).
28. Hartsock, "The Feminist Standpoint," 291–92.

But there are important differences between the perspectives available from the activities of the male worker and of women. Women's "double-day" means that a greater proportion of their lives is spent in this kind of work. Furthermore, "women also produce/reproduce men (and other women) on both a daily and long-term basis." This work requires a different kind of "production process"—transforming "natural objects" into cultural ones—from men's typical kinds of labor: "The female experience of bearing and rearing children involves a unity of mind and body more profound than is possible in the worker's instrumental activity." Women's work processes children, food, all bodies, balky machines, and social relations. It makes possible men's retreat to and appropriation of "abstract masculinity."[29]

Starting our research from women's activities in these gender divisions of labor enables us to understand how and why social and cultural phenomena have taken the forms in which they appear to us. Women's transformation of natural objects into cultural ones remains invisible, as a social activity, to men. More objective research requires restoring to our vision as necessary human social activity these "lost" processes and their relation to the activities centered in men's discourses.

(7) Women, and especially women researchers, are "outsiders within." Sociologist Patricia Hill Collins has developed feminist standpoint theory to explain the important contributions that African American feminist scholars can make to sociology—and, I would add, to our understanding of nature and social life more generally: "As outsiders within, Black feminist scholars may be one of many distinct groups of marginal intellectuals whose standpoints promise to enrich sociological discourse. Bringing this group, as well as those who share an outsider within status vis-a-vis sociology, into the center of analysis may reveal views of reality obscured by more orthodox approaches."[30] It is not enough to be only on the "outside"—to be immersed only in "women's work" or in "black women's work"—because the relations between this work and "ruling work" are not visible from only one side of this division of human activity. Instead, it is when one works on both sides that there emerges the possibility of seeing the relation between dominant activities and beliefs and those that arise on the

29. Ibid., 293, 294, 296.
30. Collins, "Learning from the Outsider Within," S15.

"outside." Bell Hooks captures this point in the title of her book *Feminist Theory: From Margin to Center.*[31] The strangers and outsiders discussed in the older anthropological and sociological writings were, consciously or not, assumed to be members of the dominant or "center" culture who were observing the residents in the dominated or marginalized cultures. No one expected the "natives" to write books about the anthropologists or sociologists (let alone be expected to sit on their tenure and promotion committees). Yet "studying up" and "studying oneself" as an "outsider within" offer resources for decreasing the partiality and distortion of research additional to those available to researchers who restrict their work to "studying down."

Dorothy Smith develops this ground when she points out in a geological metaphor that for women sociologists (may I add "women researchers" more generally?) a "line of fault" opens up between their experiences of their lives and the dominant conceptual schemes and that it is this disjuncture along which much of the major work in the women's movement has focused, especially centering on issues about women's bodies and violence against women. So objectivity is increased by thinking out of the gap between the lives of "outsiders" and the lives of "insiders" and their favored conceptual schemes.

(8) This is the right time in history. A final reason for the greater adequacy of research that begins with women's lives is suggested by parallels between feminist standpoint theories and Marxist discussions of the "standpoint of the proletariat."[32] It was not possible to see the class system of bourgeoisie and proletariat until the mid-nineteenth century, Engels argued. Utopian socialists such as Charles Fourier and Robert Owens could see the unnecessary misery and excessive wealth created at opposite ends of this emerging class system at the turn of the nineteenth century, but they could not identify in capitalism the mechanism that was producing these two classes from the peasants, artisans, merchants, and aristocrats that preceded them. The problem was not that the utopians were lacking in intellectual brilliance or that they were victims of false social myths; the reason they could not produce an adequate causal account was that the class system had not yet appeared in forms that made such explanations possible: "The great thinkers of

31. Bell Hooks, *Feminist Theory: From Margin to Center* (Boston: South End Press, 1983).
32. I examined this parallel in "Why Has the Sex/Gender System Become Visible Only Now?"

the Eighteenth Century could, no more than their predecessors, go beyond the limits imposed upon them by their epoch," observed Engels. Only with the emergence of a "conflict between productive forces and modes of production"—a conflict that "exists, in fact, objectively, outside us, independently of the will and actions even of the men that have brought it on"—could the class structure of earlier societies be detected for the first time. "Modern socialism is nothing but the reflex, in thought, of this conflict in fact; its ideal reflection in the minds, first, of the class directly suffering under it, the working class."[33]

Similarly, the sex/gender system appeared as a possible object of knowledge only with various recent changes in the situation of women and men—changes created by shifts in the economy, by the so-called sexual revolution, by the increased entrance of women into higher education, by the civil rights struggles of the 1960s, and by other identifiable economic, political, and social phenomena. The cumulative result is that the social order generates conflicting demands on and expectations for women in each and every class. Looking at nature and social relations from the perspective of these conflicts in the sex/gender system—in our lives and in other women's lives—has enabled feminist researchers to provide empirically and theoretically better accounts than can be generated from the perspective of the dominant ideology, which cannot see these conflicts and contradictions as clues to the possibility of better explanations of nature and social life.[34]

Comments

Several comments are in order before I proceed to evaluate standpoint epistemology as I did feminist empiricism. First, note that none of the foregoing claims suggests that the biological differences between women and men provide the resources for feminist analyses. Nor do these accounts appeal to women's intuition.

Second, the eight claims should be understood not as competing but as complementary ways to describe these resources. Nor should they be thought to constitute a complete list of the resources to be gained by

33. Friedrich Engels, "Socialism: Utopian and Scientific," in *The Marx and Engels Reader,* ed. Richard Tucker (New York: Norton, 1972), 606, 624.

34. For one illuminating analysis of such contradictions, see Natalie Sokoloff, "Motherwork and Working Mothers," in *Feminist Frameworks,* ed. Alison M. Jaggar and Paula S. Rothenberg (New York: McGraw-Hill, 1978).

basing research in women's lives. Feminist thinkers have identified others. For example, literary critics write of what happens when "the Other" gazes back insolently at "the self" who is the assumedly invisible agent or author of Western thought, instead of dropping her gaze demurely as "Others" are supposed to do. Women stand in the position of "the Other" to men of the dominant groups. Psychoanalytic theorists offer resources here, too, when they point out a woman is the first model for "the Other" from which the infant comes to separate its "self." And we could discuss whether the perspective from women's lives is as conducive as the perspective from the lives of men in the dominant group to assumptions that the world is "out there," ready for reflecting in our mirrorlike minds, or whether it is not more easily apparent that language is never a transparent medium and that the world-as-object-of-knowledge is and will always remain socially constructed.

I have presented the foregoing list of claims to suggest the multiple and diverse grounds upon which feminist standpoint theory rests. I also wish to counter the tendency of some postmodernist critics to insist that feminist standpoint theory is irretrievably mired in essentialist claims about mothering or some other particular activity of (some) women (see Chapter 7). The facts of mother work (for some particular cultural group) were called on in some of the original accounts to make the then not so obvious point that women's lives *were and are* different from men's in scientifically and epistemologically significant ways.

Third, I must stress that these standpoint approaches enable one to appropriate and redefine objectivity. In a hierarchically organized society, objectivity cannot be defined as requiring (or even desiring) value-neutrality (see Chapter 6).

Virtues

Standpoint epistemologies are most convincing to thinkers who are used to investigating the relationship between patterns of thought and the historical conditions that make such patterns reasonable. Consequently, many historians, political theorists, and sociologists of knowledge can find these explanations of why feminist research can generate improved research results more plausible than feminist empiricism.

The diversity of the resources that in other forms are familiar in the social sciences, and that feminists can call on in defending the greater objectivity attainable by starting research from women's lives, is another great advantage. It is hard to imagine how to defeat this entire collection of arguments—and the others to be found in feminist research—since they are grounded in a variety of relatively conventional understandings in the social sciences.

Moreover, the standpoint theories, like feminist empiricism, can claim historical precedents. Many (though not necessarily all) of the grounds identified above are used by the new histories of science to explain the emergence of modern science.[35] Scientific method itself was created by a "new kind of person" in the early modern era. Feudalism's economic order separated hand and head labor so severely that neither serfs nor aristocrats could get the necessary combination of a trained intellect and willingness to get one's hands dirty that are necessary for experimental method. One can also point to pre-Newtonian science's involvement in political struggles against the aristocracy. Or one can focus on the "fit" of Ptolemaic astronomy's conceptual scheme with the hierarchical social structure of the Catholic Church and feudal society while, in contrast, the Copernican astronomy mirrored the more democratic social order that was emerging. Or one can note the way the problematics of the new physics were "for" the rise of the new merchant classes: it was not that Newton set out to "conspire" with these classes; rather, his new physics solved problems that had to be solved if transportation, mining, and warfare were to be more efficient.[36] So the feminist empiricists' appeals to historical precedent can be made in a different way by the standpoint theorists.

35. A good overview for modern science is Wolfgang Van den Daele, "The Social Construction of Science," in *The Social Production of Scientific Knowledge,* ed. Everett Mendelsohn, Peter Weingart, and Richard Whitley (Dordrecht: Reidel, 1977).

36. See, e.g., Boris Hessen, *The Economic Roots of Newton's Principia* (New York: Howard Fertig, 1971); Edgar Zilsel, "The Sociological Roots of Science," *American Journal of Sociology* 47 (1942). A historical precedent of a different sort is claimed by Marxist theorist Fredric Jameson, who argues that although it was the Hungarian Marxist Georg Lukács who was responsible for the original development of standpoint theory, it is not Lukács's defenders today or other contemporary Marxists but feminist standpoint theorists who now exhibit "the most authentic descendency of Lukács' thinking." See Jameson, *"History and Class Consciousness* as an 'Unfinished Project,' "* in *Rethinking Marxism* 1:1 (1988), 49–72; Lukács, *History and Class Consciousness* (Cambridge, Mass.: MIT Press, 1971).

Beyond the Paternal Discourse

Like feminist empiricism, feminist standpoint theory reveals key problems in its parental discourse. Whereas Marxism states that sexism is entirely a consequence of class relations, a problem deriving only from superstructural social institutions and bourgeois ideology, in the feminist version sex/gender relations are at least as causal as economic relations in creating forms of social life and belief. Like feminist empiricism, the standpoint approach—in contrast to Marxist assumptions—takes women and men to be fundamentally sex classes, not merely or perhaps even primarily members of economic classes—though class, like race and culture, does mediate our opportunities to gain empirically adequate understandings of nature and social life. Just as feminist empiricism's radical future points toward epistemological assumptions that empiricism cannot accommodate, so the feminist standpoint's radicalism points toward epistemological assumptions that Marxism cannot contain.[37]

Clearly, it is premature to attempt now a full summary evaluation of this new tendency in feminist thought. A full assessment of the benefits of standpoint theory will have to await an exploration of the stronger criteria for objectivity that it appears to require, a clearer location of it in the conventionally divided terrain of "philosophy versus sociology," responses to skeptical questions from postmodernists and postcolonial critics, and further reflection on the relation between experience and knowledge that it advances. These are central issues for the following chapters. Nevertheless, we can make some useful preliminary assessments at this point.

Should one have to choose between feminist empiricism and the feminist standpoint as justificatory strategies? There are many projects for which feminist standpoint theory is more satisfactory, but there are at least a few for which it is not. As noted above, a justificatory strategy is intended to convince, and it is important to notice that these two are likely to appeal to different audiences. Feminist empiricism is useful precisely because it stresses the continuities between conventional justifications of scientific research and feminist ones as these would most often be understood by natural and social scientists. Feminist standpoint theory, by contrast, stresses the continuities between

37. Chapter 7 and Part III develop further the logic of feminist standpoint theory, making the contrast with its Marxist origins even more striking.

broad social formations and characteristic patterns of belief, which can be appreciated by historians, political theorists, sociologists of knowledge. These kinds of arguments will be familiar to those who know the post-Kuhnian histories and sociologies of science. A main theme of this book is the importance of conducting scientific and political arguments in the places where scientific and political decisions are being made, not only to the already convinced or only in social contexts that have minimal effect on policy-making. Many different social and political cultures influence the condition of women's lives; scientific and political decisions are made in many different places in society, and feminist research activities can find reasonable justifications in all of them.

An additional reason to avoid choosing one of these theories of knowledge over the other is that in certain respects they appear to be locked into dialogue (whether or not this relationship is always recognized by its participants), reflecting the struggles in mainstream discourses between liberal and Marxist theories of human nature and politics.[38] Perhaps choosing one over the other risks choosing more than feminism should want of those paternal discourses; we are shaped by what we reject as well as by what we accept.

These two are not the only feminist epistemologies that have been developed, but they are the two main ones to have emerged from reflection on feminist research in the natural and social sciences.[39] A third, postmodernism, has developed largely in opposition to both science and the epistemology-centered philosophy favored by societies (such as modern Western ones) that highly value scientific rationality. Modern science and epistemology-centered thought are linked, and postmodernism—including feminist postmodernism—is opposed to them. I address feminist postmodernist issues more directly in later chapters, after exploring further the "strong objectivity" that the logic of standpoint epistemologies appears to require.

38. See Jaggar, *Feminist Politics.*
39. Some critics of feminist epistemology (including some feminist critics) have claimed that there is a feminist epistemology which holds that women's experiences ground feminist knowledge claims. Some critics think that this is what feminist standpoint theory holds. Sometimes this purported epistemology is referred to as gynocentric epistemology or "female-centered" epistemology. The articulation of women's experiences does play an important role in feminist epistemologies; as I argue in Chapters 6 and 11, articulating women's experiences does make possible less partial and distorted knowledge, but it does not provide knowledge with firm foundations—it does not ground it.

6

"Strong Objectivity"
and Socially Situated Knowledge

In the preceding chapter I argued that a feminist standpoint theory can direct the production of less partial and less distorted beliefs. This kind of scientific process will not merely acknowledge the social-situatedness—the historicity—of the very best beliefs any culture has arrived at or could in principle "discover" but will use this fact as a resource for generating those beliefs.[1] Nevertheless, it still might be thought that this association of objectivity with socially situated knowledge is an impossible combination. Has feminist standpoint theory really abandoned objectivity and embraced relativism? Or, alternatively, has it remained too firmly entrenched in a destructive objectivism that increasingly is criticized from many quarters?

The Declining Status of "Objectivism"

Scientists and science theorists working in many different disciplinary and policy projects have objected to the conventional notion of a value-free, impartial, dispassionate objectivity that is supposed to guide scientific research and without which, according to conventional thought, one cannot separate justified belief from mere opinion, or real knowledge from mere claims to knowledge. From the perspective of this conventional notion of objectivity—sometimes referred to as "objec-

1. See Donna Haraway, "Situated Knowledges: *The Science Question in Feminism* and the Privilege of Partial Perspective," *Feminist Studies* 14:3 (1988).

tivism"—it has appeared that if one gives up this concept, the only alternative is not just a cultural relativism (the sociological assertion that what is thought to be a reasonable claim in one society or sub-culture is not thought to be so in another) but, worse, a judgmental or epistemological relativism that denies the possibility of any reasonable standards for adjudicating between competing claims. Some fear that to give up the possibility of one universally and eternally valid standard of judgment is perhaps even to be left with no way to argue rationally against the possibility that *each person's* judgment about the regularities of nature and their underlying causal tendencies must be regarded as equally valid. The reduction of the critic's position to such an absurdity provides a powerful incentive to question no further the conventional idea that objectivity requires value-neutrality. From the perspective of objectivism, judgmental relativism appears to be the only alternative.

Insistence on this division of epistemological stances between those that firmly support value-free objectivity and those that support judg-mental relativism—a dichotomy that unfortunately has gained the consent of many critics of objectivism as well as its defenders—has succeeded in making value-free objectivity look much more attractive to natural and social scientists than it should. It also makes judgmental relativism appear far more progressive than it is. Some critics of the conventional notion of objectivity have openly welcomed judgmental relativism.[2] Others have been willing to tolerate it as the cost they think they must pay for admitting the practical ineffectualness, the proliferation of confusing conceptual contradictions, and the political regressiveness that follow from trying to achieve an objectivity that has been defined in terms of value-neutrality. But even if embracing judg-mental relativism could make sense in anthropology and other social sciences, it appears absurd as an epistemological stance in physics or biology. What would it mean to assert that no reasonable standards can or could in principle be found for adjudicating between one culture's claim that the earth is flat and another culture's claim that the earth is round?

The literature on these topics from the 1970s and 1980s alone is huge and located in many disciplines. Prior to the 1960s the issue was

---

2. See, e.g., David Bloor, *Knowledge and Social Imagery* (London: Routledge & Kegan Paul, 1977); and many of the papers in *Knowledge and Reflexivity,* ed. Steve Woolgar (Beverly Hills, Calif.: Sage, 1988).

primarily one of ethical and cultural absolutism versus relativism. It was the concern primarily of philosophers and anthropologists and was considered relevant only to the social sciences, not the natural sciences. But since then, the recognition has emerged that cognitive, scientific, and epistemic absolutism are both implicated in ethical and cultural issues and are also independently problematic. One incentive to the expansion was Thomas Kuhn's account of how the natural sciences have developed in response to what scientists have found "interesting," together with the subsequent post-Kuhnian philosophy and social studies of the natural sciences.[3] Another has been the widely recognized failure of the social sciences to ground themselves in methods and theoretical commitments that can share in the scientificity of the natural sciences. Paradoxically, the more "scientific" social research becomes, the less objective it becomes.[4]

Further incentives have been such political tendencies as the U.S. civil rights movement, the rise of the women's movement, the decentering of the West and criticisms of Eurocentrism in international circles, and the increasing prominence within U.S. political and intellectual life of the voices of women and of African Americans and other people of Third World descent. From these perspectives, it appears increasingly arrogant for defenders of the West's intellectual traditions to continue to dismiss the scientific and epistemological stances of Others as caused mainly by biological inferiority, ignorance, underdevelopment, primitiveness, and the like. On the other hand, although diversity, pluralism, relativism, and difference have their valuable political and intellectual uses, embracing them resolves the political-scientific-epistemological conflict to almost no one's satisfaction.

I make no attempt here to summarize the arguments of these numerous and diverse writings.[5] My concern is more narrowly focused: to

3. Thomas Kuhn, *The Structure of Scientific Revolutions* (Chicago: University of Chicago Press, 1962).

4. This is an important theme in Richard Bernstein, *Beyond Objectivism and Relativism* (Philadelphia: University of Pennsylvania Press, 1983). Similar doubts about the ability of legal notions of objectivity to advance justice appear in many of the essays in "Women in Legal Education: Pedagogy, Law, Theory, and Practice," *Journal of Legal Education* 38 (1988), special issue, ed. Carrie Menkel-Meadow, Martha Minow, and David Vernon.

5. Discussions on one or more of these focuses can be found in Martin Hollis and Steven Lukes, eds., *Rationality and Relativism* (Cambridge, Mass: Harvard University Press, 1982); Michael Krausz and Jack Meiland, eds., *Relativism: Cognitive and Moral*

state as clearly as possible how issues of objectivity and relativism appear from the perspective of a feminist standpoint theory.

Feminist critics of science and the standpoint theorists especially have been interpreted as supporting either an excessive commitment to value-free objectivity or, alternatively, the abandonment of objectivity in favor of relativism. Because there are clear commitments within feminism to tell less partial and distorted stories about women, men, nature, and social relations, some critics have assumed that feminism must be committed to value-neutral objectivity. Like other feminists, however, the standpoint theorists have also criticized conventional sciences for their arrogance in assuming that they could tell one true story about a world that is out there, ready-made for their reporting, without listening to women's accounts or being aware that accounts of nature and social relations have been constructed within men's control of gender relations. Moreover, feminist thought and politics as a whole are continually revising the ways they bring women's voices and the perspectives from women's lives to knowledge-seeking, and they are full of conflicts between the claims made by different groups of feminists. How could feminists in good conscience do anything but abandon any agenda to legitimate one over another of these perspectives? Many feminists in literature, the arts, and the humanities are even more resistant than those in the natural and social sciences to claims that feminist images or representations of the world hold any special epistemological or scientific status. Such policing of thought is exactly what they have objected to in criticizing the authority of their disciplinary canons on the grounds that such authority has had the effect of stifling the voices of marginalized groups. In ignoring these views, feminist epistemologists who are concerned with natural or social science agendas appear to support an epistemological divide between the

(Notre Dame, Ind.: University of Notre Dame Press, 1982); Richard Bernstein, *Beyond Objectivism;* and S. P. Mohanty, "Us and Them: On the Philosophical Bases of Political Criticism," *Yale Journal of Criticism* 2:2 (1989). A good brief bibliographic essay on the recent philosophy of science within and against which the particular discussion of this chapter is located is Steve Fuller, "The Philosophy of Science since Kuhn: Readings on the Revolution That Has Yet to Come," *Choice,* December 1989. For more extended studies that are not incompatible with my arguments here, see Steve Fuller, *Social Epistemology* (Bloomington: Indiana University Press, 1988); and Joseph Rouse, *Knowledge and Power: Toward a Political Philosophy of Science* (Ithaca: Cornell University Press, 1987).

sciences and humanities, a divide that feminism has elsewhere criticized.

The arguments of this book move away from the fruitless and depressing choice between value-netural objectivity and judgmental relativism. The last chapter stressed the greater objectivity that can be and has been claimed to result from grounding research in women's lives. This chapter draws on some assumptions underlying the analyses of earlier chapters in order to argue that the conventional notion of objectivity against which feminist criticisms have been raised should be regarded as excessively weak. A feminist standpoint epistemology requires strengthened standards of objectivity. The standpoint epistemologies call for recognition of a historical or sociological or cultural relativism—but not for a judgmental or epistemological relativism. They call for the acknowledgment that all human beliefs—including our best scientific beliefs—are socially situated, but they also require a critical evaluation to determine which social situations tend to generate the most objective knowledge claims. They require, as judgmental relativism does not, a scientific account of the relationships between historically located belief and maximally objective belief. So they demand what I shall call *strong objectivity* in contrast to the weak objectivity of objectivism and its mirror-linked twin, judgmental relativism. This may appear to be circular reasoning—to call for scientifically examining the social location of scientific claims—but if so, it is at least not viciously circular.[6]

This chapter also considers two possible objections to the argument presented, one that may arise from scientists and philosophers of science, and another that may arise among feminist themselves.

6. Additional writings informing this chapter include esp. Haraway, "Situated Knowledges"; Donna Haraway, *Primate Visions: Gender, Race, and Nature in the World of Modern Science* (New York: Routledge, 1989); Jane Flax, *Thinking Fragments: Psychoanalysis, Feminism, and Postmodernism in the Contemporary West* (Berkeley: University of California Press, 1990); and the writings of standpoint theorists themselves, esp. Nancy Hartsock, "The Feminist Standpoint: Developing the Ground for a Specifically Feminist Historical Materialism," in *Discovering Reality: Feminist Perspectives on Epistemology, Metaphysics, Methodology, and Philosophy of Science*, ed. Sandra Harding and Merrill Hintikka (Dordrecht: Reidel, 1983); Dorothy Smith, *The Everyday World as Problematic: A Feminist Sociology* (Boston: Northeastern University Press, 1987); Hilary Rose, "Hand, Brain, and Heart: A Feminist Epistemology for the Natural Sciences," *Signs* 9:1 (1983); Patricia Hill Collins, "Learning from the Outsider Within: The Sociological Significance of Black Feminist Thought," *Social Problems* 33 (1986)—though each of these theorists would no doubt disagree with various aspects of my argument.

Objectivism's Weak Conception of Objectivity

The term "objectivism" is useful for the purposes of my argument because its echoes of "scientism" draw attention to ways in which the research prescriptions called for by a value-free objectivity only mimic the purported style of the most successful scientific practices without managing to produce their effects. Objectivism results only in semiscience when it turns away from the task of critically identifying all those broad, historical social desires, interests, and values that have shaped the agendas, contents, and results of the sciences much as they shape the rest of human affairs. Objectivism encourages only a partial and distorted explanation of why the great moments in the history of the natural and social sciences have occurred.

Let me be more precise in identifying the weaknesses of this notion. It has been conceptualized both too narrowly and too broadly to be able to accomplish the goals that its defenders claim it is intended to satisfy. Taken at face value it is ineffectively conceptualized, but this is what makes the sciences that adopt weak standards of objectivity so effective socially: objectivist justifications of science are useful to dominant groups that, consciously or not, do not really intend to "play fair" anyway. Its internally contradictory character gives it a kind of flexibility and adaptability that would be unavailable to a coherently characterized notion.

Consider, first, how objectivism operationalizes too narrowly the notion of maximizing objectivity. The conception of value-free, impartial, dispassionate research is supposed to direct the identification of all social values and their elimination from the results of research, yet it has been operationalized to identify and eliminate *only* those social values and interests that differ among the researchers and critics who are regarded by the scientific community as competent to make such judgments. If the community of "qualified" researchers and critics systematically excludes, for example, all African Americans and women of all races, and if the larger culture is stratified by race and gender and lacks powerful critiques of this stratification, it is not plausible to imagine that racist and sexist interests and values would be identified within a community of scientists composed entirely of people who benefit—intentionally or not—from institutional racism and sexism.

This kind of blindness is advanced by the conventional belief that the truly scientific part of knowledge-seeking—the part controlled by methods of research—is only in the context of justification. The context of discovery, where problems are identified as appropriate for scientific investigation, hypotheses are formulated, key concepts are defined—this part of the scientific process is thought to be unexaminable within science by rational methods. Thus "real science" is restricted to those processes controllable by methodological rules. The methods of science—or, rather, of the special sciences—are restricted to procedures for the testing of already formulated hypotheses. Untouched by these careful methods are those values and interests entrenched in the very statement of what problem is to be researched and in the concepts favored in the hypotheses that are to be tested. Recent histories of science are full of cases in which broad social assumptions stood little chance of identification or elimination through the very best research procedures of the day.[7] Thus objectivism operationalizes the notion of objectivity in much too narrow a way to permit the achievement of the value-free research that is supposed to be its outcome.

But objectivism also conceptualizes the desired value-neutrality of objectivity too broadly. Objectivists claim that objectivity requires the elimination of *all* social values and interests from the research process and the results of research. It is clear, however, that not all social values and interests have the same bad effects upon the results of research. Some have systematically generated less partial and distorted beliefs than others—or than purportedly value-free research—as earlier chapters have argued.

Nor is this so outlandish an understanding of the history of science as objectivists frequently intimate. Setting the scene for his study of nineteenth-century biological determinism, Stephen Jay Gould says:

> I do not intend to contrast evil determinists who stray from the path of scientific objectivity with enlightened antideterminists who approach

7. This is the theme of many feminist, left, and antiracist analyses of biology and social sciences. See, e.g., Anne Fausto-Sterling, *Myths of Gender: Biological Theories about Women and Men* (New York: Basic Books, 1985); Stephen Jay Gould, *The Mismeasure of Man* (New York: Norton, 1981); Robert V. Guthrie, *Even the Rat Was White: A Historical View of Psychology* (New York: Harper & Row, 1976); Haraway, *Primate Visions;* Sandra Harding, ed., *Feminism and Methodology: Social Science Issues* (Bloomington: Indiana University Press, 1987); Joyce Ladner, ed., *The Death of White Sociology* (New York: Random House, 1973); Hilary Rose and Steven Rose, eds., *Ideology of/in the Natural Sciences* (Cambridge, Mass.: Schenkman, 1979); Londa Schiebinger, *The Mind Has No Sex: Women in the Origins of Modern Science* (Cambridge, Mass.: Harvard University Press, 1989).

data with an open mind and therefore see truth. Rather, I criticize the myth that science itself is an objective enterprise, done properly only when scientists can shuck the constraints of their culture and view the world as it really is. . . . Science, since people must do it, is a socially embedded activity. It progresses by hunch, vision, and intuition. Much of its change through time does not record a closer approach to absolute truth, but the alteration of cultural contexts that influence it so strongly.[8]

Other historians agree with Gould.[9] Modern science has again and again been reconstructed by a set of interests and values—distinctively Western, bourgeois, and patriarchal—which were originally formulated by a new social group that intentionally used the new sciences in their struggles against the Catholic Church and feudal state. These interests and values had both positive and negative consequences for the development of the sciences.[10] Political and social interests are not "add-ons" to an otherwise transcendental science that is inherently indifferent to human society; scientific beliefs, practices, institutions, histories, and problematics are constituted in and through contemporary political and social projects, and always have been. It would be far more startling to discover a kind of human knowledge-seeking whose products could—alone among all human products—defy historical "gravity" and fly off the earth, escaping entirely their historical location. Such a cultural phenomenon would be cause for scientific alarm; it would appear to defy principles of "material" causality upon which the possibility of scientific activity itself is based.[11]

Of course, people in different societies arrive at many of the same empirical claims. Farmers, toolmakers, and child tenders in every culture must arrive at similar "facts" about nature and social relations if their work is to succeed. Many of the observations collected by medieval European astronomers are preserved in the data used by

---

8. Gould, *Mismeasure of Man,* 21–22.

9. E.g., William Leiss, *The Domination of Nature* (Boston: Beacon Press, 1972); Carolyn Merchant, *The Death of Nature: Women, Ecology, and the Scientific Revolution* (New York: Harper & Row, 1980); Wolfgang Van den Daele, "The Social Construction of Science," in *The Social Production of Scientific Knowledge,* ed. Everett Mendelsohn, Peter Weingart, and Richard Whitley (Dordrecht: Reidel, 1977).

10. The usefulness of such political movements to the growth of knowledge in the sciences is discussed in Chapter 3.

11. See Chapter 4. Rouse, *Knowledge and Power,* provides a good analysis of the implications for science of Foucauldian notions of politics and power.

astronomers today. But what "facts" these data refer to, what further research they point to, what theoretical statements they support and how such theories are to be applied, what such data signify in terms of human social relations and relations to nature—all these parts of the sciences can differ wildly, as the contrast between medieval and contemporary astronomy illustrates.

There are yet deeper ways in which political values permeate modern science. For even relatively conservative tendencies in the post-Kuhnian philosophies of science, the sciences' power to manipulate the world is considered the mark of their success. The "new empiricism" contrasts in this respect with conventional empiricism. As Joseph Rouse puts the point:

> If we take the new empiricism seriously, it forces us to reappraise the relation between power and knowledge in a more radical way. The central issue is no longer how scientific claims can be distorted or suppressed by polemic, propaganda, or ideology. Rather, we must look at what was earlier described as the achievement of power through the application of knowledge. But the new empiricism also challenges the adequacy of this description in terms of "application." The received view distinguishes the achievement of knowledge from its subsequent application, from which this kind of power is supposed to derive. New empiricist accounts of science make this distinction less tenable by shifting the locus of knowledge from accurate representation to successful manipulation and control of events. Power is no longer external to knowledge or opposed to it: power itself becomes the mark of knowledge.[12]

The best as well as the worst of the history of the natural sciences has been shaped by—or, more accurately, constructed through and within—political desires, interests, and values. Consequently, there appear to be no grounds left from which to defend the claim that the objectivity of research is advanced by the elimination of all political values and interests from the research process. Instead, the sciences need to legitimate *within scientific research,* as part of practicing science, crit-

---

12. Rouse, *Knowledge and Power,* 19. Among the "new empiricist" works that Rouse has in mind are Larry Laudan, *Progress and Its Problems: Toward a Theory of Scientific Growth* (Berkeley: University of California Press, 1977); Mary Hesse, *Revolutions and Reconstructions in the Philosophy of Science* (Bloomington: University of Indiana Press, 1980); Nancy Cartwright, *How the Laws of Physics Lie* (Oxford: Oxford University Press, 1983).

ical examination of historical values and interests that may be so shared within the scientific community, so invested in by the very constitution of this or that field of study, that they will not show up as a cultural bias between experimenters or between research communities. What objectivism cannot conceptualize is the need for critical examination of the "intentionality of nature"—meaning not that nature is no different from humans (in having intentions, desires, interests, and values or in constructing its own meaningful "way of life," and so on) but that nature as-the-object-of-human-knowledge never comes to us "naked"; it comes only as already constituted in social thought.[13] Nature-as-object-of-study simulates in this respect an intentional being. This idea helps counter the intuitively seductive idea that scientific claims are and should be an epiphenomenon of nature. It is the development of strategies to generate just such critical examination that the notion of strong objectivity calls for.

Not everyone will welcome such a project; even those who share these criticisms of objectivism may think the call for strong objectivity too idealistic, too utopian, not realistic enough. But is it more unrealistic than trying to explain the regularities of nature and their underlying causal tendencies scientifically but refusing to examine *all* their causes? And even if the ideal of identifying all the causes of human beliefs is rarely if ever achievable, why not hold it as a desirable standard? Anti-litter laws improve social life even if they are not always obeyed.[14]

Weak objectivity, then, is a contradictory notion, and its contradictory character is largely responsible for its usefulness and its widespread appeal to dominant groups. It offers hope that scientists and science institutions, themselves admittedly historically located, can produce claims that will be regarded as objectively valid without their having to examine critically their own historical commitments, from which—intentionally or not—they actively construct their scientific research. It permits scientists and science institutions to be unconcerned with the origins or consequences of their problematics and practices, or with the social values and interests that these problematics and practices support. It offers the possibility of enacting what Francis

13. See Haraway, *Primate Visions*, esp. chap. 10, for analysis of differences between the Anglo-American, Japanese, and Indian constructions of "nature" which shape the objects of study in primatology.

14. Fuller uses the anti-litter law example in another context in *Social Epistemology.*

Bacon promised: "The course I propose for the discovery of sciences is such as leaves but little to the acuteness and strength of wits, but places all wits and understandings nearly on a level." His "way of discovering sciences goes far to level men's wits, and leaves but little to individual excellence; because it performs everything by surest rules and demonstrations."[15]

For those powerful forces in society that want to appropriate science and knowledge for their own purposes, it is extremely valuable to be able to support the idea that ignoring the constitution of science within political desires, values, and interests will somehow increase the reliability of accounts of nature and social life. The ideal of the disinterested rational scientist advances the self-interest of both social elites and, ironically, scientists who seek status and power. Reporting on various field studies of scientific work, Steve Fuller points out that Machiavellian judgments

> simulate those of the fabled "rational" scientist, since in order for the Machiavellian to maximize his advantage he must be ready to switch research programs when he detects a change in the balance of credibility—which is, after all, what philosophers of science would typically have the rational scientist do. To put the point more strikingly, it would seem that as the scientist's motivation approximates total *self-interestedness* (such that he is always able to distance his own interests from those of any social group which supports what may turn out to be a research program with diminishing credibility), his behavior approximates total *disinterestedness*. And so we can imagine the ultimate Machiavellian scientist pursuing a line of research frowned upon by most groups in the society—perhaps determining the racial component in intelligence is an example—simply because he knows of its potential for influencing the course of future research and hence for enhancing his credibility as a scientist.[16]

The history of science shows that research directed by maximally liberatory social interests and values tends to be better equipped to identify partial claims and distorting assumptions, even though the credibility of the scientists who do it may not be enhanced during the short run. After all, antiliberatory interests and values are invested in the natural inferiority of just the groups of humans who, if given real equal access

15. Quoted in Van den Daele, "Social Construction of Science," 34.
16. Fuller, *Social Epistemology,* 267.

(not just the formally equal access that is liberalism's goal) to public voice, would most strongly contest claims about their purported natural inferiority. Antiliberatory interests and values silence and destroy the most likely sources of evidence against their own claims. That is what makes them rational for elites.

Strong Objectivity: A Competency Concept

At this point, what I mean by a concept of strong objectivity should be clear. In an important sense, our cultures have agendas and make assumptions that we as individuals cannot easily detect. Theoretically unmediated experience, that aspect of a group's or an individual's experience in which cultural influences cannot be detected, functions as part of the evidence for scientific claims. Cultural agendas and assumptions are part of the background assumptions and auxiliary hypotheses that philosophers have identified. If the goal is to make available for critical scrutiny *all* the evidence marshaled for or against a scientific hypothesis, then this evidence too requires critical examination *within* scientific research processes. In other words, we can think of strong objectivity as extending the notion of scientific research to include systematic examination of such powerful background beliefs. It must do so in order to be competent at maximizing objectivity.

The strong objectivity that standpoint theory requires is like the "strong programme" in the sociology of knowledge in that it directs us to provide symmetrical accounts of both "good" and "bad" belief formation and legitimation.[17] We must be able to identify the social causes of good beliefs, not just of the bad ones to which the conventional "sociology of error" and objectivism restrict causal accounts. However, in contrast to the "strong programme," standpoint theory requires causal analyses not just of the micro processes in the laboratory but also of the macro tendencies in the social order, which shape scientific practices. Moreover, a concern with macro tendencies permits a more robust notion of reflexivity than is currently available in the sociology of knowledge or the philosophy of science. In trying to identify the social causes of good beliefs, we will be led also to examine

---

17. I use "good" and "bad" here to stand for "true" and "false," "better confirmed" and "less well confirmed," "plausible" and "implausible," and so on.

critically the kinds of bad beliefs that shape our own thought and behaviors, not just the thought and behavior of others.

To summarize the argument of the last chapter, in a society structured by gender hierarchy, "starting thought from women's lives" increases the objectivity of the results of research by bringing scientific observation and the perception of the need for explanation to bear on assumptions and practices that appear natural or unremarkable from the perspective of the lives of men in the dominant groups. Thinking from the perspective of women's lives makes strange what had appeared familiar, which is the beginning of any scientific inquiry.[18]

Why is this gender difference a scientific resource? It leads us to ask questions about nature and social relations from the perspective of devalued and neglected lives. Doing so begins research in the perspective from the lives of "strangers" who have been excluded from the culture's ways of socializing the "natives," who are at home in its institutions and who are full-fledged citizens. It starts research in the perspective from the lives of the systematically oppressed, exploited, and dominated, those who have fewer interests in ignorance about how the social order actually works. It begins research in the perspective from the lives of people on the "other side" of gender battles, offering a view different from the "winner's stories" about nature and social life which men's interpretations of men's lives tend to produce. It starts thought in everyday life, for which women are assigned primary responsibility and in which appear consequences of dominant group activities—consequences that are invisible from the perspective of those activities. It starts thought in the lives of those people to whom is assigned the work of mediating many of the culture's ideological dualisms—especially the gap between nature and culture. It starts research in the lives not just of strangers or outsiders but of "outsiders within," from which the relationship between outside and inside, margin and center, can more easily be detected. It starts thought in the perspective from the life of the Other, allowing the Other to gaze back "shamelessly" at the self who had reserved for himself the right to gaze "anonymously" at whomsoever he chooses. It starts thought in the lives of people who are unlikely to permit the denial of the interpretive

---

18. As emphasized in Chapters 5 and 7, starting thought from women's lives is something that both men and women must *learn* to do. Women's telling their experiences is not the same thing as thinking from the perspective of women's lives.

core of all knowledge claims. It starts thought in the perspective from lives that at this moment in history are especially revealing of broad social contradictions. And no doubt there are additional ways in which thinking from the perspective of women's lives is especially revealing of regularities in nature and social relations and their underlying causal tendencies.

As analyzed further in Part III, it is important to remember that in a certain sense there are no "women" or "men" in the world—there is no "gender"—but only women, men, and gender constructed through particular historical struggles over just which races, classes, sexualities, cultures, religious groups, and so forth, will have access to resources and power. Moreover, standpoint theories of knowledge, whether or not they are articulated as such, have been advanced by thinkers concerned not only with gender and class hierarchy (recollect that standpoint theory originated in class analyses) but also with other "Others."[19] To make sense of any actual woman's life or the gender relations in any culture, analyses must begin in real, historic women's lives, and these will be women of particular races, classes, cultures, and sexualities. The historical particularity of women's lives is a problem for narcissistic or arrogant accounts that attempt, consciously or not, to conduct a cultural monologue. But it is a resource for those who think that our understandings and explanations are improved by what we could call an intellectual participatory democracy.

The notion of strong objectivity welds together the strengths of weak objectivity and those of the "weak subjectivity" that is its correlate, but excludes the features that make them only weak. To enact or operationalize the directive of strong objectivity is to value the Other's perspective and to pass over in thought into the social condition that creates it—not in order to stay there, to "go native" or merge the self with the Other, but in order to look back at the self in all its cultural particularity from a more distant, critical, objectifying location. One

---

19. See, e.g., Samir Amin, *Eurocentrism* (New York: Monthly Review Press, 1989); Bettina Aptheker, *Tapestries of Life: Women's Work, Women's Consciousness, and the Meaning of Daily Life* (Amherst: University of Massachusetts Press, 1989); Collins, "Learning from the Outsider Within"; Walter Rodney, *How Europe Underdeveloped Africa* (Washington, D.C.: Howard University Press, 1982); Edward Said, *Orientalism* (New York: Pantheon Books, 1978); Edward Said, Foreword to *Selected Subaltern Studies,* ed. Ranajit Guha and Gayatri Chakravorty Spivak (New York: Oxford University Press, 1988), viii.

can think of the subjectivism that objectivism conceptualizes as its sole alternative as only a "premodern" alternative to objectivism; it provides only a premodern solution to the problem *we* have here and now at the moment of postmodern criticisms of modernity's objectivism. Strong objectivity rejects attempts to resuscitate those organic, occult, "participating consciousness" relationships between self and Other which are characteristic of the premodern world.[20] Strong objectivity requires that we investigate the relation between subject and object rather than deny the existence of, or seek unilateral control over, this relation.

Historical Relativism versus Judgmental Relativism

It is not that historical relativism is in itself a bad thing. A respect for historical (or sociological or cultural) relativism is always useful in starting one's thinking. Different social groups tend to have different patterns of practice and belief and different standards for judging them; these practices, beliefs, and standards can be explained by different historical interests, values, and agendas. Appreciation of these empirical regularities are especially important at this moment of unusually deep and extensive social change, when even preconceived schemes used in liberatory projects are likely to exclude less-well-positioned voices and to distort emerging ways of thinking that do not fit easily into older schemes. Listening carefully to different voices and attending thoughtfully to others' values and interests can enlarge our vision and begin to correct for inevitable enthnocentrisms. (The dominant values, interests, and voices are not among these "different" ones; they are the powerful tide against which "difference" must swim.)

To acknowledge this historical or sociological fact, as I have already argued, does not commit one to the further epistemological claim that there are therefore no rational or scientific grounds for making judgments between various patterns of belief and their originating social practices, values, and consequences. Many thinkers have pointed out that judgmental relativism is internally related to objectivism. For ex-

---

20. See Morris Berman, *The Reenchantment of the World* (Ithaca: Cornell University Press, 1981), for an analysis of the world that modernity lost, and lost for good. Some feminists have tried to dismantle modernist projects with premodernist tools.

ample, science historian Donna Haraway argues that judgmental relativism is the other side of the very same coin from "the God trick"
required by what I have called weak objectivity. To insist that no
judgments at all of cognitive adequacy can legitimately be made
amounts to the same thing as to insist that knowledge can be produced
only from "no place at all": that is, by someone who can be every place
at once.[21] Critical preoccupation with judgmental relativism is the
logical complement to the judgmental absolutism characteristic of Eurocentrism. Economist Samir Amin criticizes the preoccupation with
relativism in some Western intellectual circles as a kind of "inverted
Eurocentrism":

> The view that any person has the right—and even the power—to judge
> others is replaced by attention to the relativity of those judgments.
> Without a doubt, such judgments can be erroneous, superficial, hasty,
> or relative. No case is ever definitely closed; debate always continues.
> But that is precisely the point. It is necessary to pursue debate and not
> to avoid it on the grounds that the views that anyone forms about
> others are and always will be false: that the French will never under-
> stand the Chinese (and vice versa), that men will never understand
> women, etc; or, in other words, that there is no human species, but only
> "people." Instead, the claim is made that only Europeans can truly
> understand Europe, Chinese China, Christians Christianity, and
> Moslems Islam; the Eurocentrism of one group is completed by the
> inverted Eurocentrism of others.[22]

Historically, relativism appears as a problematic intellectual possibility only for dominating groups at the point where the hegemony of
their views is being challenged. Though the recognition that other
cultures do, in fact, hold different belief, values, and standards of
judgment is as old as human history, judgmental relativism emerged as
an urgent intellectual issue only in nineteenth-century Europe, with the
belated recognition that the apparently bizarre beliefs and behaviors of
Others had a rationality and logic of their own. Judgmental relativism

21. Haraway, "Situated Knowledges" makes these points and uses the phrase "the
God trick."
22. Amin, *Eurocentrism*, 146–47. Amin further makes clear that it takes more than
mere debate—i.e., only intellectual work—to come to understand the lives or point of
view of "people" who are on trajectories that oppose one's own in political struggles.
The following paragraph draws on "Introduction: Is There a Feminist Method?" in
Feminism and Methodology, p. 10.

is not a problem originating in or justifiable in terms of the lives of marginalized groups. It did not arise in misogynous thought about women; it does not arise from the contrast feminism makes between women's lives and men's. Women do not have the problem of how to accommodate intellectually both the sexist claim that women are inferior in some way or another and the feminist claim that they are not. Here relativism arises as a problem only from the perspective of men's lives. Some men want to appear to acknowledge and accept feminist arguments without actually giving up any of their conventional androcentric beliefs and the practices that seem to follow so reasonably from such beliefs. "It's all relative, my dear," is a convenient way to try to accomplish these two goals.

We feminists in higher education may have appeared to invite charges of relativism in our language about disseminating the results of feminist research and scholarship beyond women's studies programs into the entire curriculum and canon. We speak of "mainstreaming" and "integrating" the research, scholarship, and curriculum of Other programs and of encouraging "inclusiveness" in scholarship and the curriculum. We enroll our women's studies courses in campuswide projects to promote "cultural diversity" and "multiculturalism," and we accept students into such courses on these terms. Do these projects conflict with the standpoint logic? Yes and no. They conflict because the notions involved are perfectly coherent with the maintenance of elitist knowledge production and systems. Let me make the point in terms of my racial identity as white. "They (those people of color at the margins of the social order) are to be integrated with us (whites at the center), leaving us unchanged and the rightful heirs of the center of the culture. They are to give up their agendas and interests that conflict with ours in order to insert their contributions into the research, scholarship, or curriculum that has been structured to accommodate our agendas and interests." This is just as arrogant a posture as the older cultural absolutism. From the perspective of racial minorities, integration has never worked as a solution to ethnic or race relations in the United States. Why is there reason to think it will work any better for the marginalized projects in intellectual circles?

Should we therefore give up attempts at an "inclusive curriculum" and "cultural diversity" because of their possible complicity with sexism, racism, Eurocentrism, heterosexism, and class oppression? Of course the answer must be no. It is true that this kind of language

appears to betray the compelling insights of the standpoint epistemology and to leave feminist programs in the compromised position
of supporting the continued centering of white, Western, patriarchal
visions. But many feminist projects—including women's studies programs themselves—are forced to occupy whatever niches they can find
within institutional structures that are fundamentally opposed to them
or, at least, "prefeminist." An implicit acceptance of pluralism, if not
judgmental relativism—at least at the institutional level—appears to
be the only condition under which women's voices and feminist voices,
male and female, can be heard at all.

After all, isn't feminism just one "equal voice" among many competing for everyone's attention? The nineteenth-century "natives"
whose beliefs and behaviors Europeans found bizarre were not in any
real sense competing for an equal voice within European thought and
politics. They were safely off in Africa, the Orient, and other faraway
places. The chances were low that aborigines would arrive in Paris,
London, and Berlin to study and report back to their own cultures the
bizarre beliefs and behaviors that constituted the "tribal life" of European anthropologists and *their* culture. More important, there was no
risk at all that they could have used such knowledge to assist in impos-
ing their rule on Europeans in Europe. Women's voices, while certainly
far from silent, were far more effectively contained and muted than is
possible today. As a value, a moral prescription, relativism was a safe
stance for Europeans to choose; the reciprocity of respect it appeared
to support had little chance of having to be enacted. Today, women and
feminists are not safely off and out of sight at all. They are present,
speaking, within the very social order that still treats women's beliefs
and behaviors as bizarre. Moreover, their speech competes for attention and status as most plausible not only with that of misogynists but
also with the speech of other Others: African Americans, other peoples
of color, gay rights activists, pacifists, ecologists, members of new
formations of the left, and so on. Isn't feminism forced to embrace
relativism by its condition of being just one among many countercultural voices?

This description of the terrain in which feminists struggle to advance
their claims, however, assumes that people must either choose only one
among these countercultures as providing an absolute standard for
sorting knowledge claims, or else regard all of them as competing and
assign them equal cognitive status. Actually, it is a different scenario

that the countercultures can envision and even occasionally already enact: the fundamental tendencies of each must permeate each of the others in order for each movement to succeed. Feminism should center the concerns of each of these movements, and each of them must move feminist concerns to its center.

To summarize, then, a strong notion of objectivity requires a commitment to acknowledge the historical character of every belief or set of beliefs—a commitment to cultural, sociological, historical relativism. But it also requires that judgmental or epistemological relativism be rejected. Weak objectivity is located in a conceptual interdependency that includes (weak) subjectivity and judgmental relativism. One cannot simply give up weak objectivity without making adjustments throughout the rest of this epistemological system.

Responding to Objections

Two possible objections to the recommendation of a stronger standard for objectivity must be considered here. First, some scientists and philosophers of science may protest that I am attempting to specify standards of objectivity for all the sciences. What could it mean to attempt to specify *general* standards for increasing the objectivity of research? Shouldn't the task of determining what counts as adequate research be settled within each science by its own practitioners? Why should practicing scientists revise their research practices because of what is thought by a philosopher or anyone else who is not an expert in a particular science?

But the issue of this chapter is an epistemological issue—a meta-scientific one—rather than an issue within any single science. It is more like a directive to operationalize theoretical concepts than like a directive to operationalize in a certain way some particular theoretical notion within physics or biology. The recommended combination of strong objectivity with the acknowledgment of historical relativism would, if adopted, create a culturewide shift in the kind of epistemology regarded as desirable. Certainly, strategies for enacting commitments to strong objectivity and the acknowledgment of historical relativism would have to be developed within each particular research program; plenty of examples already exist in biology and the social sciences. My position is that the natural sciences are backward in this

respect; they are not immune from the reasonableness of these directives, as conventionalists have assumed.

The notion of strong objectivity developed here represents insights
that have been emerging from thinkers in a number of disciplines for
some decades—not just "wishful thinking" based on no empirical
sciences at all. Criticisms of the dominant thought of the West from
both inside and outside the West argue that its partiality and distortions are the consequence in large part of starting that thought only
from the lives of the dominant groups in the West. Less partiality and
less distortion result when thought starts from peasant life, not just
aristocratic life; from slaves' lives, not just slaveowners' lives; from the
lives of factory workers, not just those of their bosses and managers;
from the lives of people who work for wages and have also been
assigned responsibility for husband and child care, not just those of
persons who are expected to have little such responsibility. This directive leaves open to be determined within each discipline or research
area what a researcher must do to start thought from women's lives or
the lives of people in other marginalized groups, and it will be easier—
though still difficult—to provide reasonable responses to such a request in history or sociology than in physics or chemistry. But the
difficulty of providing an analysis in physics or chemistry does not
signify that the question is an absurd one for knowledge-seeking in
general, or that there are no reasonable answers for those sciences too.

The second objection may come from feminists themselves. Many
would say that the notion of objectivity is so hopelessly tainted by its
historical complicity in justifying the service of science to the dominant
groups that trying to make it function effectively and progressively in
alternative agendas only confuses the matter. If feminists want to
breathe new life into such a bedraggled notion as objectivity, why not
at least invent an alternative term that does not call up the offenses
associated with the idea of value-neutrality, that is not intimately tied
to a faulty theory of representation, to a faulty psychic construction of
the ideal agent of knowledge, and to regressive political tendencies.

Let us reorganize some points made earlier in order to get the full
force of this objection. The goal of producing results of research that
are value-free is part of the notion of the ideal mind as a mirror that can
reflect a world that is "out there," ready-made (see Chapter 4). In this
view, value-free objectivity can locate an Archimedean perspective
from which the events and processes of the natural world appear in

their proper places. Only false beliefs have social causes—human values and interests that blind us to the real regularities and underlying causal tendencies in the world, generating biased results of research. True beliefs have only natural causes: those regularities and underlying causal tendencies that are *there,* plus the power of the eyes to see them and of the mind to reason about them. This theory of representation is a historically situated one: it is characteristic only of certain groups in the modern West. Can the notion of objectivity really be separated from this implausible theory of representation?

Value-free objectivity requires also a faulty theory of the ideal agent—the subject—of science, knowledge, and history. It requires a notion of the self as a fortress that must be defended against polluting influences from its social surroundings. The self whose mind would perfectly reflect the world must create and constantly police the borders of a gulf, a no-man's-land, between himself as the subject and the object of his research, knowledge, or action. Feminists have been among the most pointed critics of this self-versus-Other construct,[23] referring to it as "abstract masculinity."[24] Moreover, its implication in Western constructions of the racial Other against which the "white" West would define its admirable projects is also obvious.[25] Can the notion of objectivity be useful in efforts to oppose such sexism and racism?

Equally important, the notion of value-free objectivity is morally and politically regressive for reasons additional to those already mentioned. It justifies the construction of science institutions and individual scientists as "fast guns for hire." It has been used to legitimate and hold up as the highest ideal institutions and individuals that are, insofar as they are scientific, to be studiously unconcerned with the

23. See, e.g., Nancy Chodorow, *The Reproduction of Mothering* (Berkeley: University of California Press, 1978); Dorothy Dinnerstein, *The Mermaid and the Minotaur: Sexual Arrangements and Human Malaise* (New York: Harper & Row, 1976); Carol Gilligan, *In a Different Voice: Psychological Theory and Women's Development* (Cambridge, Mass.: Harvard University Press, 1982); Evelyn Fox Keller, *Reflections on Gender and Science* (New Haven, Conn.: Yale University Press, 1984).
24. Hartsock, "The Feminist Standpoint."
25. See, e.g., Sander Gilman, *Difference and Pathology: Stereotypes of Sexuality, Race, and Madness* (Ithaca: Cornell University Press, 1985); V. Y. Mudimbe, *The Invention of Africa: Gnosis, Philosophy, and the Order of Knowledge* (Bloomington: Indiana University Press, 1988); Said, *Orientalism,* and Foreword to Guha and Spivak, *Subaltern Studies.*

origins or consequences of their activities or with the values and interests that these activities advance. This nonaccidental, determined,
energetic lack of concern is supported by science education that excludes training in critical thought and that treats all expressions of
social and political concern—the concerns of the torturer and the
concerns of the tortured—as being on the same low level of scientific
"rationality." Scandalous examples of the institutional impotence of
the sciences as sciences to speak to the moral and political issues that
shape their problematics, consequences, values, and interests have
been identified for decades (see Chapter 4). The construction of a
border between scientific method and violations of human and, increasingly, animal rights must be conducted "outside" that method, by
government statements about what constitutes acceptable methods of
research on human and animal subjects, what constitutes consent to
experimentation, the subsequent formation of "ethics committees,"
and so on. Can the notion of objectivity be extracted from the morals
and politics of "objective science" as a "fast gun for hire"?

These are formidable objections. Nevertheless, the argument of this
book is that the notion of objectivity not only can but should be
separated from its shameful and damaging history. Research is socially
situated, and it can be more objectively conducted without aiming for
or claiming to be value-free. The requirements for achieving strong
objectivity permit one to abandon notions of perfect, mirrorlike representations of the world, the self as a defended fortress, and the "truly
scientific" as disinterested with regard to morals and politics, yet still
apply rational standards to sorting less from more partial and distorted belief. Indeed, my argument is that these standards are more
rational and more effective at producing maximally objective results
than the ones associated with what I have called weak objectivity.

As I have been arguing, objectivity is one of a complex of inextricably linked notions. Science and rationality are two other terms in this
network. But it is not necessary to accept the idea that there is only one
correct or reasonable way to think about these terms, let alone that the
correct way is the one used by dominant groups in the modern West.
Not all reason is white, masculinist, modern, heterosexual, Western
reason. Not all modes of rigorous empirical knowledge-seeking are
what the dominant groups think of as science—to understate the
point. The procedures institutionalized in conventional science for distinguishing between how we want the world to be and how it is are not

the only or best ways to go about maximizing objectivity. It is important to work and think outside the dominant modes, as the minority movements have done. But it is important, also, to bring the insights developed there into the heart of conventional institutions, to disrupt the dominant practices from within by appropriating notions such as objectivity, reason, and science in ways that stand a chance of compelling reasoned assent while simultaneously shifting and displacing the meanings and referents of the discussion in ways that improve it. It is by thinking and acting as "outsiders within" that feminists and others can transform science and its social relations for those who remain only insiders or outsiders.

One cannot afford to "just say no" to objectivity. I think there are three additional good reasons to retain the notion of objectivity for future knowledge-seeking projects but to work at separating it from its damaging historical associations with value-neutrality.

First, it has a valuable political history. There have to be standards for distinguishing between how I want the world to be and how, in empirical fact, it is. Otherwise, might makes right in knowledge-seeking just as it tends to do in morals and politics. The notion of objectivity is useful because its meaning and history support such standards. Today, as in the past, there are powerful interests ranged against attempts to find out the regularities and underlying causal tendencies in the natural and social worlds. Some groups do not want exposed to public scrutiny the effect on the environment of agribusiness or of pesticide use in domestic gardening. Some do not want discussed the consequences for Third World peasants, for the black underclass in the United States, and especially for women in both groups of the insistence on economic production that generates profit for elites in the West. The notion of achieving greater objectivity has been useful in the past and can be today in struggles over holding people and institutions responsible for the fit between their behavior and the claims they make.

Second, objectivity also can claim a glorious intellectual history. The argument of this chapter has emphasized its service to elites, but it also has been invoked to justify unpopular criticisms of partisan but entrenched beliefs. Standpoint theory can rightfully claim that history as its legacy.

Finally, the appeal to objectivity is an issue not only between feminist and prefeminist sciences but within each feminist and other emancipatory movement. There are many feminisms, some of which result in claims that distort the racial, class, sexuality, and gender relationships in society. Which ones generate less and which more partial and distorted accounts of nature and social life? The notion of objectivity is useful in providing a way to think about the gap we want between how any individual or group wants the world to be and how in fact it is.

The notion of objectivity—like such ideas as science and rationality, democracy and feminism—contains progressive as well as regressive tendencies. In each case, it is important to develop the progressive and to block the regressive ones.

Reflexivity Revisited

The notion of "strong objectivity" conceptualizes the value of putting the subject or agent of knowledge in the same critical, causal plane as the object of her or his inquiry. It permits us to see the scientific as well as the moral and political advantages of this way of trying to achieve a reciprocal relationship between the agent and object of knowledge. The contrast developed here between weak and strong notions of objectivity permits the parallel construction of weak versus strong notions of reflexivity.

Reflexivity has tended to be seen as a problem in the social sciences—and only there. Observation cannot be as separated from its social consequences as the directives of "weak objectivity," originating in the natural sciences, have assumed. In social inquiry, observation changes the field observed. Having recognized his complicity in the lives of his objects of study, the researcher is then supposed to devise various strategies to try to democratize the situation, to inform the "natives" of their options, to make them participants in the account of their activities, and so forth.[26]

26. A fine account of the travails of such a project reports Robert Blauner and David Wellman's dawning recognition that nothing they did could eliminate the colonial relationship between themselves and their black informants in the community surrounding Berkeley; see their "Toward the Decolonization of Social Research," in Ladner, *The Death of White Sociology*. Economist Vernon Dixon argues that from the perspective of an African or African American world view, the idea that observation would not change the thing observed appears ridiculous; see his "World Views and Research Methodology," in *African Philosophy: Assumptions and Paradigms for Research on Black Persons*, ed. L. M. King, Vernon Dixon, and W. W. Nobles (Los Angeles: Fanon Center,

Less commonly, reflexivity has been seen as a problem because if the researcher is under the obligation to identify the social causes of the "best" as well as the "worst" beliefs and behaviors of those he studies, then he must also analyze his own beliefs and behaviors in conducting his research project—which have been shaped by the same kinds of social relations that he is interested to identify as causes of the beliefs and behaviors of others. (Here, reflexivity can begin to be conceptualized as a "problem" for the natural sciences, too.) Sociologists of knowledge in the recent "strong programme" school and related tendencies, who emphasize the importance of identifying the social causes of "best belief," have been aware of this problem from the very beginning but have devised no plausible way of resolving it—primarily because their conception of the social causes of belief in the natural sciences (the subject matter of their analyses) is artificially restricted to the micro processes of the laboratory and research community, explicitly excluding race, gender, and class relations. This restricted notion of what constitutes appropriate subject matter for analyses of the social relations of the sciences is carried into their understanding of their own work. It generates ethnographies of their own and the natural science communities which are complicitous with positivist tendencies in insisting on the isolation of research communities from the larger social, economic, and political currents in their societies. (These accounts are also flawed by their positivist conceptions of the object of natural science study).[27]

These "weak" notions of reflexivity are disabled by their lack of any mechanism for identifying the cultural values and interests of the researchers, which form part of the evidence for the results of research in both the natural and social sciences. Anthropologists, sociologists, and the like, who work within social communities, frequently appear to desire such a mechanism or standard; but the methodological assumptions of their disciplines, which direct them to embrace either weak objectivity or judgmental relativism, have not permitted them to develop one. That is, individuals express "heartfelt desire" not to harm the

---

Charles R. Drew Postgraduate Medical School, 1976), and my discussion of the congruence between African and feminine world views in *The Science Question in Feminism* (Ithaca: Cornell University Press, 1986), chap. 7.

27. See, e.g., Bloor, *Knowledge and Social Imagery;* and Steve Woolgar's nevertheless interesting paper, "Reflexivity Is the Ethnographer of the Text," as well as other (somewhat bizarre) discussions of reflexivity in Woolgar, *Knowledge and Reflexivity.*

subjects they observe, to become aware of their own cultural biases, and so on, but such reflexive goals remain at the level of desire rather than competent enactment. In short, such weak reflexivity has no possible operationalization, or no competency standard, for success.

A notion of strong reflexivity would require that the objects of inquiry be conceptualized as gazing back in all their cultural particularity and that the researcher, through theory and methods, stand behind them, gazing back at his own socially situated research project in all its cultural particularity and its relationships to other projects of his culture—many of which (policy development in international relations, for example, or industrial expansion) can be seen only from locations far away from the scientist's actual daily work.[28] "Strong reflexivity" requires the development of oppositional theory from the perspective of the lives of those Others ("nature" as already socially constructed, as well as other peoples), since intuitive experience, for reasons discussed earlier, is frequently not a reliable guide to the regularities of nature and social life and their underlying causal tendencies.

Standpoint theory opens the way to stronger standards of both objectivity and reflexivity. These standards require that research projects use their historical location as a resource for obtaining greater objectivity.

28. This notion is developed more fully in Chapter 11.

7

Feminist Epistemology
in and after the Enlightenment

Chapter 5 examined the outlines of the two leading theories of knowledge that have been developed in response to skeptical questions about the possibility of feminist contributions to the natural and social sciences: feminist empiricism, and feminist standpoint theory. Chapter 6 responded to the charge that the second of these abandons objectivity and embraces a damaging kind of relativism, arguing that on the contrary the logic of standpoint theory calls for *stronger* standards of objectivity than those conventionally adopted, that it avoids damaging forms of relativism, and that it transforms the reflexivity of research from a problem into a scientific resource.

The standpoint approaches must be defended not only against conventionalists, however, but also against an array of more innovative theorists who think standpoint epistemology far too conservative. Criticisms of standpoint theory often conflict. On the one hand, is standpoint theory really an epistemology, or only a sociology of knowledge? On the other hand, does it claim foundations for knowledge just at the moment when such claims are being criticized from many quarters? That is, is it too epistemological and not sociological enough? Many conventionalists think that it devalues science; other critics argue that it values science too highly. Is it essentialist, grounding its arguments in claims to some universal notion of "woman" or, at least, of women's activity? Is it rightly the target of various criticisms by postmodernists, or is it—can it be—part of postmodernism? This chapter attempts to answer such questions and then to examine further the links between the standpoint theory and feminist postmodernism.[1]

Contradictory Criticisms

Standpoint theorists get caught in the current debates between three groups of intellectuals: epistemologists and philosophers of science, sociologists of knowledge, and postmodernist critics. For partially different reasons, the sociologists and postmodernists think that the standpoint approach is too epistemological, while—again for partially different reasons—the philosophers and the very same postmodernists think that it is too sociological. Perhaps these disputes reflect, in part, competing attempts by literary and cultural critics, social scientists, and philosophers to increase the intellectual and professional status of their disciplines. Even if this is true, however, it is also the case that these disputes attempt to work through the difficult issue of what to save and what to abandon of the conceptual frameworks of humanism and the Enlightenment.[2] We can sort through at least some of the issues in this complex discussion by responding to five charges from the various critics of standpoint theory.

(1) Is standpoint theory excessively foundationalist and thus "too epistemological"? Recent accounts by sociologists and historians of science have criticized the idea that knowledge can have any certain or transsocially firm foundations. The search for clear and certain foundations has generated the agendas of philosophers from Plato and Aristotle through John Locke, George Berkeley, David Hume, Immanuel Kant, Bertrand Russell, A. J. Ayer, and Roderick Chisholm. The

1. I first took up the topic of the relationship between feminism and postmodernism in *The Science Question in Feminism* (Ithaca: Cornell University Press, 1986) and then in "Feminism, Science, and the Anti-Enlightenment Critiques," in *Feminism/Postmodernism*, ed. Linda J. Nicholson (New York: Routledge, 1989). My arguments here diverge from those in the earlier discussions.

2. The two agendas are not incompatible, because humanism and the Enlightenment have the effect of valuing disciplines in specific historical ways at the expense of other disciplines and other values. Neither "humanism" or "the Enlightenment," however, is really the right term for what is being evaluated in these discussions, nor is "the modern" (versus postmodern) or "the scientific world view." No one has suggested a completely unproblematic term.

reasonableness of such agendas depends upon assuming that only false beliefs have social causes. As long as true beliefs—or, at least, beliefs not yet known to be false—are assumed not to have social causes, there is conceptual space for conventional philosophers to debate a logic of belief that appears to them to require no historical or sociological assumptions or "inputs"; conventional epistemology can claim a terrain that it takes to be completely distinct from the terrain of sociology. But this territory disappears, as the "strong programme" sociologists point out, once we understand that our "best beliefs" too have social causes.[3] If that is true, conventional epistemology appears to have no distinctive field of inquiry.

Sociologists who adopt a "strong programme" properly find in its domain the task of producing descriptions and explanations of the social causes of true belief, of knowledge. Conventional sociology of knowledge, which these theorists refer to as the "weak programme," permitted the production of descriptions and explanations only of false beliefs and of "knowers." It provided a sociology of error and a sociology of knowers but not really a sociology of *knowledge* at all. It could explain the social causes of the rise of Ptolemaic astronomy, phlogiston theory, or Lysenkoism but saw nothing to be explained— no sociological subject matter—in the social acceptance of Copernican astronomy, Newtonian mechanics, or evolutionary theory; when observation of and reason about the regularities of nature were the only subject matters, explanations of our "best beliefs" had to be left to the philosphers and, of course, scientists. The "strong programme" theorists argue, however, that a fully scientific sociology of knowledge will seek to replace mystical philosophical accounts with scientific histories and sociologies of how some beliefs get legitimated and others do not. The same kinds of causes that are cited to explain the generation and acceptance of false beliefs must be used to explain the generation and acceptance of "true beliefs"; what they refer to as "causal symmetry" should be the goal of maximally scientific accounts of belief acceptance. From this perspective, they assert, the feminist standpoint theorists are trying to reinstate foundations to knowledge just when the

3. David Bloor, *Knowledge and Social Imagery* (London: Routledge & Kegan Paul, 1977); Barry Barnes, *Interests and the Growth of Knowledge* (Boston: Routledge & Kegan Paul, 1977).

effort to do so is finally becoming discredited in the social studies of science.[4]

One response to such criticism is that the kinds of foundations claimed by standpoint theorists do not much look like the proposals for foundations that conventional epistemologists formulate; the perspective from women's lives does not much resemble the sense data statements, such as "Here now red field," claimed as foundations for knowledge by Bertrand Russell and other philosophers. Nevertheless, the standpoint theorists do argue that women's lives provide scientifically preferable starting points for generating and testing scientific hypotheses compared with the lives of men in the dominant groups, which have conventionally though unofficially been used to generate scientific problems and against which knowledge claims have disproportionately been tested. It is because of this argument that feminist standpoint theory is said to be reinstating foundationalism.

However, standpoint theorists are not claiming some kind of transhistorical privilege for research that begins in women's lives. Nor, it is worth recollecting, are they arguing that women's biology, women's intuition, what women say, or women's experiences provide grounds for knowledge. What women say and what women experience do provide important clues for research designs and results, but it is the objective perspective *from women's lives* that gives legitimacy to feminist knowledge, according to standpoint theorists.[5]

A variety of analyses in the prefeminist literature have the inadvertent effect of supporting feminist standpoint theory; we need not invent new arguments to defeat criticism of foundationalism. Philosopher Charles Mills, in his defense of such alternative epistemologies as feminist standpoint theory, draws on such an analysis:

As several critics have argued, one can accept symmetry about the *fact* of causation [of belief] while still rejecting it with respect to the *nature* of causation, and its probable differential consequences. W. H. Newton-Smith contrasts the cases of two people with particular beliefs about where they are sitting; only one of these people has operative

---

4. "Strong programme," sociology of knowledge is flawed in a number of ways, but I consider here only the kinds of criticism of standpoint theory that it generates.

5. Chapter 5 expounded this distinction. Chapter 11 develops the argument that men can and must learn how to generate original knowledge with the assistance of perspectives from womens' lives.

perceptual facilities. In both cases, belief is the result of causal processes, but this symmetry does not extend deeper: "In the case of a veridical perceptual belief the causal chain involved runs through the state of affairs that gives the belief its truth-value. With non-veridical perceptual beliefs the causal chain may have nothing to do with the state of affairs that gives the belief its truth-value." In a parallel fashion, then, it can be argued that in the cases cited above it is the actual state of affairs which (differentially perceived) gives rise to the beliefs in particular social groups. Once we allow reasons to be causes, there is no contradiction in affirming that beliefs can be simultaneously socially and rationally caused.[6]

Thus, "strong programme" defenders should not be able to defeat the feminist standpoint claims that one can rationally distinguish social conditions giving rise to false beliefs from those giving rise to less false ones. We should be skeptical about claims to have identified transhistorical grounds for knowledge, but this skepticism does not require us to refuse the possibility of historical causes (as well as natural ones) for knowledge claims. We should be able to provide sociological accounts of the development and social acceptance of Copernican astronomy, Newton's mechanics, and evolutionary theory without thereby undermining their claims to provide knowledge. These scientific beliefs, like those that feminist standpoint theories justify, are held for *both* good reasons and social causes.

There is another problem with the "strong programme" accounts and others like them.[7] They are flawed by their inability to provide a reasonable account of the social causes of their own production and the high value rightly accorded to so many of their insights. It is not just that these accounts happen to fall into relativism; David Bloor, for example, actively embraces it. Although he notes that his commitment to value-neutrality makes his program relativist and unable to progress beyond descriptive sociological accounts of how appeals to objectivity function as resources in science, he does not regard this limitation as a

6. Charles Mills, "Alternative Epistemologies," *Social Theory and Practice* 14:3 (1988), 251. The quotation is from W. H. Newton-Smith, *The Rationality of Science* (Boston: Routledge & Kegan Paul, 1981), 253. Mills directs the reader also to Warren Schmaus, "Reasons, Causes, and the 'Strong Programme' in the Sociology of Knowledge," *Philosophy of the Social Sciences* 15 (1985).

7. E.g., Bruno Latour and Steve Woolgar, *Laboratory Life: The Social Construction of Scientific Facts* (Beverly Hills, Calif.: Sage, 1979); Karin Knorr-Cetina, *The Manufacture of Knowledge* (Oxford: Pergamon Press, 1981).

serious problem.[8] But as noted in Chapter 6, standpoint theory can insist on placing the observer, and institutions of observation such as science, in the same critical plane as what is observed. The same kinds of social forces that shape the rest of the world shape our own accounts, including scientific ones.

This is a reason to get our inquiry processes and institutions inserted into the kinds of social contexts that have tended to cause less false rather than more false beliefs. This is a reason to think critically about how beliefs are formed and why they are adopted, and to develop such thought as a *logic* of belief that carries normative values, not just as "scientific" description of what in fact occurs. It is as naive as the epistemological policies that the "strong programme" theorists criticize to adopt a policy of refusing to infer logics of belief from sociological and historical descriptions of belief formation and acceptance. Standpoint theory claims that we can provide good reasons for dividing beliefs into the false and the probably less false (or "bad" and "good" beliefs) but that these reasons do not refer to transcendental, certain grounds for belief of the sort claimed by conventional epistemologies or privilege what any group of actual, historical humans say about how they see the world (as misreadings of the Marxist epistemology have assumed).

(2) Are the standpoint epistemologies not epistemologies at all but really only sociologies of knowledge? The preceding point enables us to address more quickly the opposing charge from philosophers that these purported philosophies of knowledge in fact have no epistemological content at all. Here is how philosopher Charles Mills explains that they are indeed epistemologies, even though their concerns are different from conventional ones. Standpoint theorists

do not, for the most part, see themselves as offering, within the conventional framework, alternative analyses of such traditional epistemological topics as memory, perception, truth, belief, and so on, or coming up with startling new solutions to the Gettier problem. Nor is their paradigmatic cognizer that familiar Cartesian figure, the abstract, disembodied, individual knower, beset by skeptical and solipsist hazards, trying to establish a reliable cognitive relationship with the basic furniture of the Universe. . . .

Characteristically, then, their concerns will be not the problem of

8. Bloor, *Knowledge and Social Imagery,* 141–44.

other minds, but the problem of why women were not thought to have minds; not an investigation of the conditions under which individual memory is reliable, but an investigation of the social conditions under which systematic historical amnesia about the achievements of African civilizations became possible; not puzzlement about whether or not physical objects exist, but puzzlement about the cognitive mechanisms which make relational social properties appear under capitalism as reified intrinsic natural properties.[9]

It is easy for conventional philosophers to overlook the fact that all epistemologies make assumptions about how beliefs in fact have been generated and gained legitimacy, even though few epistemologists discuss their psychological, sociological, and historical assumptions. As Roy Bhaskar puts the point,

> any theory of knowledge presupposes a *sociology* in the sense that it must be assumed, implicitly if not explicitly, that the nature of human beings and the institutions they reproduce or transform is such that such knowledge could be produced. Thus the Humean theory presupposes a conception of people as passive sensors of given facts and recorders of their given constant conjunctions, which has the corollary that knowledge can always be analysed in a purely individualistic way.[10]

Moreover, every sociology or psychology or history must assume an epistemology, since it must assume that *its* practices do produce knowledge. Standpoint theory, like all others, contains both a sociology (history, psychology, and so on) and an epistemology. We can ask the kind of question about epistemologies that Kuhn asked about the philosophies of science: what is the point of a theory of knowledge that cannot account for the "best beliefs" that humans have produced?[11] Of course, we must *all* discuss just which beliefs are the best beliefs, but whichever they are, we want them accounted for *as* best by our theories of knowledge ("we" here should be as socially inclusive as possible, of course). When we look at the history of epistemology and philosophy of science, it is perfectly clear that the "Greats" in these fields are attempting to theorize adequately the historical changes in

9. Mills, "Alternative Epistemologies," 237, 239.
10. Bhaskar, *Reclaiming Reality,* 49–50.
11. Thomas Kuhn, *The Structure of Scientific Revolutions* (Chicago: University of Chicago Press, 1970), chap. 1.

the kinds of beliefs that their age finds reasonable and the difficulty of appealing to conventional grounds to justify them. Thus, virtually all the leading epistemologists of the modern era attempt to make sense out of the difference in the ways scientific beliefs are generated and legitimated and the consequences of these differences for other kinds of belief, such as religious or political and social belief. Descartes, Locke, Hume, Kant, and other philosophers are quite explicit about what they see as the social causes of their problematics and the new kinds of standards of adequate belief they must keep in mind. Nor is it controversial to say that the concern with epistemology and philosophy of science in the early twentieth-century Vienna Circle was caused by the fact that the most widely confirmed theory about the physical world had been rejected in favor of a theory with only the most miniscule level of empirical support: that is, Einstein's physics replaced Newton's physics. (One can give more nuanced accounts of the relationship between the two theories as well as of the motivating interests of the members of the Vienna Circle, but the foregoing summary is regarded as reasonable.)

The standpoint theories, too, take their problematic from shifts in the kinds of beliefs being generated (beliefs that challenge conventional views about women and gender), the nonconventional kinds of people producing them (women, feminists), nonconventional research processes (starting from women's rather than men's lives), and the nonconventional ways these beliefs are justified. They try to show the relationship between how more empirically and theoretically adequate beliefs are in fact being produced and the nonconventional prescriptions that are being and should be made for producing them. Thus, standpoint theories have characteristics of both conventional sociologies and conventional epistemologies. They do not confuse or conflate the two fields but explicitly articulate kinds of relationships between them that are left only implicit in the conventional disciplines.

It is not only disciplinary power but also more general political power that is at issue in attempts to segregate the philosophy and sociology of science. The claim used to justify the autonomy of science, its freedom from the checks and balances governing other social institutions that are subject to social, economic, and political influences, is that the sciences themselves are capable of keeping social forces from shaping the content of science. It is obvious however, that social, economic, and political agendas have always shaped the course of the sciences and the pictures of nature and social relations which they legitimate. Efforts to maintain a firm distinction between the philosophy and sociology of science reflect, observes Joseph Rouse, the desire to sustain "a *conceptual* separation between science viewed as a field of knowledge and science viewed as a field of power."[12] Such a strategy may hide but does not block the effects of politics on science. Standpoint theories require—they include—fully critical sociologies (and histories) of science and socially situated philosophies of science and epistemologies. Whether we *call* standpoint theories sociologies or epistemologies or philosophies of science will be settled, no doubt, by skirmishes between the disciplines. But they are, in fact, all of these.

(3) Does standpoint epistemology value science too highly? Scientists and science enthusiasts frequently object that all feminist criticisms of science are unfair, that they unjustly devalue science, its rationality, and its logic. Yet others criticize the feminist approaches to science, including the standpoint epistemology, for not separating themselves firmly enough from the view that there is or can be such a thing as pure science and thus that science is fundamentally or essentially good for society. This second criticism shares a terrain with the charge that standpoint theories are "too epistemological," but the concern of its skepticism is science rather than epistemology. According to sociologist Sal Restivo, for example, feminists do not ask how the social structure of a society is a determinant of the knowledge produced in that society:

> The sociological perspective, in the strong structural sense I argue for, is not a prominent feature of feminist science studies and criticism. This does not mean that the feminists do not draw attention to problems of social structure. They do not, however, do so in ways that transform epistemology from a philosophical to a sociological project. This makes it difficult for them to transcend the ideology of pure science. A sociological theory of knowledge must replace epistemology before we can begin to construct alternate ways of inquiry.[13]

Restivo curiously appropriates for his structural sociology the kinds of prescriptions about how to gain knowledge which have been described

---

12. Joseph Rouse, *Knowledge and Power: Toward a Political Philosophy of Science* (Ithaca: Cornell University Press, 1987), 17 (original emphasis). See also Stanley Aronowitz, *Science as Power* (Minneapolis: University of Minnesota Press, 1988).

13. Sal Restivo, "Modern Science as a Social Problem," *Social Problems* 35:3 (1988), 211, 217.

both as sociology and as epistemology by feminist standpoint theorists.[14] The final paragraph of his essay begins with the following questions: "What sorts of social formations foster disinterestedness and objectivity? That is, under what conditions can inquiry proceed unburdened as much as possible by mundane interests and commitments and within the most expansive network of information and knowledge possible?" Though he has earlier discussed modern science's complicity in the rise of capitalism and imperialist expansion, at this point the concrete details of macro politics disappear, and he presents an abstract proposal: "Based on the preceding conjectures, my answer is social formations in which the person has primacy, social formations that are diversified, cooperative, egalitarian, non-authoritarian, participatory."[15] Yes, indeed! But how we get from here to there in the face of white supremacy, new forms of imperialism, continuing class exploitation, compulsory heterosexuality, and still-powerful male supremacy remains mysterious. One would think that certain sorts of interestedness—not disinterestedness—should be linked to maximizing objectivity in the imperfect social order within which science exists today. Moreover, how science and epistemology are to contribute to this project is not asked.

Meanwhile, women want to know about how their bodies really work, just which social forces are most responsible for keeping women in poverty, why men rape, how imperialism specifically acts on women, how women can gain the power to improve their condition—and we want to know these things now, or at least as soon as possible. Knowing them is a precondition for creating the kinds of social formations valued by Restivo. That means we must investigate these scientific questions within the existing formations of our society (foundations, universities, health-care systems, the economy, and so on), which are not diversified, are uncooperative, hierarchical, authoritarian, and nonparticipatory. For this we need the kinds of "successor sciences" that the standpoint epistemologies articulate and call for.[16]

(4) Are standpoint theories essentialist? Are they Eurocentric? Wom-

14. Cf. e.g., Dorothy Smith, *The Everyday World as Problematic: A Feminist Sociology* (Boston: Northeastern University Press, 1987); and Hilary Rose, "Hand, Brain, and Heart: A Feminist Epistemology for the Natural Sciences," *Signs* 9:1 (1983).

15. Restivo, "Modern Science," 220.

16. Feminist scientific tendencies are intended to provide "successor sciences" to sexist, androcentric, class-biased, racist, and imperialist ones.

en in marginalized races, classes, sexualities, and cultures have argued that there is no such thing as "the feminine," or universal woman. Women are part of every race, class, sexuality, and culture. Their experiences, activities, struggles, and perspectives are accordingly different from one another's, not just different from some generalized "men's" lives or even only from the lives of the men in their own social group. Moreover, women differ from one another in two significantly distinctive ways: some important differences are due to cultural variation, such as the consequences of living in Puerto Rico rather than in Australia, but other differences are due to the hierarchical power relationships between women in the dominant and dominated groups. Do standpoint theories reinstate the notion that a homogeneous women's experience or activity or struggle or perspective is to serve as the grounds for feminist claims? Isn't this feminist theory just replicating one more time the racism, class bias, heterosexism, and Eurocentrism that has damaged conventional thought?[17]

This kind of criticism is reinforced by writers who for other reasons are suspicious of universalizing tendencies. Postmodernist writers argue that universal man—that socially "homogeneous," unitary, willful, and autonomous subject of Western science, reason, ethics, and history—is a construct of the Enlightenment that we are better off without. Feminism has played an important role in helping us to arrive

17. For examples of important criticisms of white, Western feminism by women of Third World descent, see Gloria Anzaldua, *Borderlands/La Frontera: The New Mestiza* (San Francisco: Spinsters/Aunt Lute, 1987); Hazel V. Carby, "White Women Listen! Black Feminism and the Boundaries of Sisterhood," in *The Empire Strikes Back: Race and Racism in 70's Britain*, ed. Center for Contemporary Cultural Studies (London: Hutchinson, 1982); Bell Hooks, *Feminist Theory: From Margin to Center* (Boston: South End Press, 1982); Gloria T. Hull, Patricia Bell Scott, and Barbara Smith, eds., *All the Women Are White, All the Blacks Are Men, but Some of Us Are Brave: Black Women's Studies* (Old Westbury, N.Y.: Feminist Press, 1982); Audre Lorde, *Sister Outsider* (Trumansburg, N.Y.: Crossing Press, 1984); Cherrie Moraga and Gloria Anzaldua, eds., *This Bridge Called My Back: Writings by Radical Women of Color* (Latham, N.Y.: Kitchen Table:Women of Color Press, 1983); Barbara Smith, ed., *Home Girls: A Black Feminist Anthology* (Latham, N.Y.: Kitchen Table:Women of Color Press, 1983); Gayatri Chakravorty Spivak, *In Other Worlds: Essays in Cultural Politics* (New York: Methuen, 1987); Maxine Baca Zinn, Lynn Weber Cannon, Elizabeth Higginbotham, and Bonnie Thornton Dill, "The Costs of Exclusionary Practices in Women's Studies," *Signs* 11:2 (1986). For criticisms of heterosexism in feminist thought, see (among others) Marilyn Frye, *The Politics of Reality: Essays in Feminist Theory* (Trumansburg, N.Y.: Crossing Press, 1983); and Adrienne Rich, "Compulsory Heterosexuality and Lesbian Existence," *Signs* 5:4 (1980).

at this conclusion, some point out, though this fact is rarely acknowledged in the nonfeminist postmodernist writings. But once universal
man began to be a questionable notion, so too did his helpmate walk-
ing three steps behind—universal woman; hence, the critics continue,
standpoint epistemology is regressive in assuming some sort of universal feminine condition that can serve as the grounds for feminist
claims. And if it does not do so, then just what is feminist about
standpoint theories? Don't their grounds deteriorate into views from
each of the thousands (millions?) of distinctive kinds of social experiences, or perspectives, or activities, or struggles characteristic of different cultural groups of women? In order to claim to be producing a
distinctively *feminist* theory of knowledge, according to these critics,
one must make faulty assumptions about universal woman, and so
standpoint theory appears doomed either way: it faces the apparently
impossible task of maneuvering between charges of Eurocentric essentialism and of the failure to provide any distinctively feminist analysis
at all. Are these charges valid?[18]

The first thing to be said is that there does appear to be a problem
with the logic that feminist standpoint theories borrow from Marxist
epistemology. Marxist theory focuses on a certain kind of difference
between the proletarian worker who sells his labor and the owner of
capital who buys it—though "difference" is a far too apolitical (or,
rather, bourgeois) term for what interested Marx in this relationship.
More marginal to the logic of Marxist analyses are national, religious,
gender, race, and other "cultural" ("superstructural") differences be-

---

18. Analyses of essentialism and Eurocentrism and consideration of these charges
appear in many discussions of feminist epistemology and, often specifically, of standpoint theory. Among the most important are Christine Di Stefano, "Dilemmas of Difference: Feminism, Modernity, and Postmodernism," in Nicholson, *Feminism/
Postmodernism;* Jane Flax, "Postmodernism and Gender Relations in Feminist Theory,"
in Nicholson, *Feminism/Postmodernism;* Jane Flax, *Thinking Fragments: Psychoanalysis, Feminism, and Postmodernism in the Contemporary West* (Berkeley: University
of California Press, 1990); Nancy Fraser and Linda J. Nicholson, "Social Criticism
without Philosophy: An Encounter between Feminism and Postmodernism," in Nicholson, *Feminism/Postmodernism;* Judith Grant, "I Feel, Therefore I Am: A Critique of
Female Experience as the Basis for a Feminist Epistemology," *Women and Politics* 7:3
(1987); Nancy Hartsock, "Epistemology and Politics: Minority vs. Majority Theories,"
Cultural Critique 7 (1987); Mary E. Hawkesworth, "Knowers, Knowing, Known: Feminist Theory and Claims of Truth," *Signs* 14:1 (1989); Susan Hekman, "The Feminization of Epistemology: Gender and the Social Sciences," *Women and Politics* 7:3 (1987);
Susan Hekman, "Comment on Hawkesworth's 'Knowers, Knowing, Known: Feminist
Theory and Claims of Truth,'" *Signs* 15:2 (1990).

tween workers. Although Marxist thinkers have commented on the special conditions of women, slaves, and Jews, and on differences between the situation of workers in Germany and in France, for example, the fundamental interest of mainstream Marxist theory and politics has been the difference created by the relationship between those who own the means of production and those who must sell their labor.[19] Similarly, feminist standpoint theory, like much other feminist theory, has been fundamentally interested in the differences created by relationships between women's and men's lives.[20]

Many readers will already have detected something peculiar about this line of thought. In the first place, the predominantly white and Western theorists who have developed the standpoint approaches have been among the most vigorous of such feminists in taking an active stance against racism, imperialism, heterosexism, and class oppression—in contrast to many feminist (and prefeminist) thinkers of other epistemological persuasions. (This should not be surprising, since the Marxist analysis on which this epistemology draws could reasonably be regarded as the most insightful and comprehensive analysis of the causes of racism and imperialism, as well as of class oppression, until recent decades. It was a valuable resource used by African Americans and Third World peoples to explain their situation.) Moreover, these same theorists have been involved in political activity alongside women of color, poor women, and lesbians—in the old socialist feminist Women's Unions, in the Jesse Jackson campaign, and in other local and national feminist, left, and antiracist activities. In the second place, it is not just white, Western, straight, economically privileged feminists who find standpoint approaches persuasive. African American theorists use them: Patricia Hill Collins draws on standpoint theory to explain what one can learn by starting inquiry in the lives of African American women. Bettina Aptheker uses it to show what one can learn about the dominant social theories if one starts from the daily lives of poor women, African American women, Japanese American women,

---

19. This is changing. One fine example of the change (which I came across just as I was finishing this chapter) is Ernesto Laclau and Chantal Mouffe's analysis in *Hegemony and Socialist Strategy: Towards a Radical Democratic Politics* (London: Verso, 1985).

20. In *The Science Question in Feminism*, I raised these skeptical questions without answering them; the result was an unduly negative impression of the resources of standpoint theory, which I attempt to rectify here.

lesbians; Samir Amin invokes it in analyzing Eurocentrism, and Edward Said in discussing Orientalism and the new histories of India.[21]

How is one to explain the apparently essentializing logic of standpoint theories in light of their clear usefulness to so many different "alternative sciences"? One possibility is that standpoint logic permits each oppressed group to center the view of the world from its members' lives while ignoring any and all significant differences between those lives and others; that is, it permits each oppressed group to ignore all other liberatory social movements' agendas. The logic of standpoint approaches does not force one to this solipsistic and essentializing stance, however, nor do I mean to suggest that the theorists mentioned above or the liberatory social movements necessarily use it in this way.

Alternatively, one can see that standpoint theory contains within it tendencies both to ignore and to emphasize differences within the groups on which it focuses—in our case, differences between women or between men. Feminist analyses do tend to slide away from focusing on differences between women as long as actual analyses from the perspective of lives of marginalized women are not specifically under way. General statements about the standpoint of women or the feminist standpoint feel as if they must be assuming gender essentialism, and some people who use the language of standpoint theory may well be essentialists. Even when one is careful not to use the term "women" to stand for all women, the logic of these arguments leads to talk of "women's experience," "women's activity," "women's oppression," "women's situation," "women's struggle," as if these events and processes were the same for all women, regardless of their race, class, or culture.

Even talking about these aspects of women's lives in the plural—"experience*s*"—does not succeed in itself in deflecting the essentializing tendency but announces an intention it has not yet taken action on: the "s" just hangs there without any analysis to support it. It is like saying "men—and women, of course" but then continuing to make the

21. Patricia Hill Collins, "Learning from the Outsider Within: The Sociological Significance of Black Feminist Thought," *Social Problems* 33 (1986); Bettina Aptheker, *Tapestries of Life: Women's Work, Women's Consciousness, and the Meaning of Daily Life* (Amherst: University of Massachusetts Press, 1989); Samir Amin, *Eurocentrism* (New York: Monthly Review Press, 1989); Edward Said, *Orientalism* (New York: Pantheon Books, 1978), and his Foreword to *Selected Subaltern Studies,* ed. Ranajit Guha and Gayatri Chakravorty Spivak (New York: Oxford University Press, 1988), viii.

same assumptions and to discuss only the same old issues that one was concerned with when one clearly meant to refer only to "male men." (Think of discussions in ethics or economics or legal theory about the "moral" or "rational" or "reasonable—uh—person.") Another symptom of the same problem is that one has to remind oneself to add a qualifier to statements about "men"—"men of the dominant races, classes, cultures"—as one recollects that working-class and minority men, however sexist and androcentric they may be, are not the men who have designed and who most benefit from the institutions and practices that are oppressive to women. (Their benefits are only a "trickle-down" effect from those of men in the ruling groups.)

Again, however, it is not that the standpoint theorists or their theories are overtly or intentionally racist or class-biased; rather, they tend to center a difference between the genders at the ontological (and, consequently, scientific, epistemological, and political) expense of clearly focusing on differences between women or between men in different races, classes, and cultures.

Nevertheless, standpoint theory also provides important resources for meeting the challenge to emphasize differences between women. It insists on the recognition that not just false claims but also true (or less false) ones are socially situated. It insists on causal symmetry in explanations of how to arrive at "good" beliefs. Thus it places the observer and her "institutions of observation" in the same critical plane as the subject matters to be observed.[22] Whatever kinds of causes are attributed to patterns in nature or social life (for example, limitations or resources that are due to gender, race, sexuality, or class relations) should also be critically examined as parts of the probable background beliefs functioning as "evidence" in those explanations. Further, the logic of analysis of the scientific advantage that accrues to starting research from the lives of the dominated gender requires acknowledgment of the similar advantage to be gained by starting from the lives of dominated groups in oppositional race, class, sexuality, or cultural relations.

But these understandings of the logic of standpoint theory must be accompanied at every point by richer conceptualizations and analyses of the interlocking relationships between sexism, racism, heterosexism,

---

22. Smith, in *Everyday World*, insists on placing the observer in the same critical plane as her subject matter; see Chapter 6.

and class oppression. Each of these phenomena is fundamentally a relation, not a "thing"; each is a dynamic relationship that constantly changes, partially because changes occur in the *other* relationships. Consequently (as noted in an earlier chapter), in an important sense it is true that in societies stratified by race, class, or culture there are no such persons as women or men per se; there are only women and men in particular, historically located race and class and cultural relations. There are no gender relations per se but only gender relations as constructed by and between classes, races, and cultures. As Sojourner Truth pointed out in her famous question "Ain't I a woman?" the femininity required of the white woman was exactly what was denied to the black slave woman.[23] Similarly, the exercises and privileges of masculinity that the white plantation owner claimed for himself were exactly what he denied the black male slave. The German historian Gisela Boch shows that the pronatalist reproductive policies legislated for the Aryan race in Germany between the world wars had their opposite in policies of enforced sterilization, abortion, and eventually genocide for Gypsies, Jews, and other groups regarded as racially inferior. She suggests the term "sexist racism" to remind us that policies or practices intended to be racially discriminatory will also in many respects be sexist.[24] The phrase also reminds us that those policies and practices which discriminate against women will usually, perhaps always, discriminate in different ways against women in different races. Similar discussions of race- and class-discriminatory reproductive policies can be found in feminist writings on the politics of reproductive policies in the United States, now and in the past, and on international reproductive policies and technologies.[25] We should think of systems of race, class, gender, and culture as interlocking: one cannot dislodge one piece without disturbing the others.[26]

It should be clear that if it is beneficial to start research, scholarship, and theory in white women's situations, then we should be able to

23. Angela Davis, *Women, Race, and Class* (New York: Random House, 1981).

24. Gisela Boch, "Racism and Sexism in Nazi Germany: Motherhood, Compulsory Sterilization, and the State," *Signs* 8:3 (1983).

25. See, e.g., Vimal Balasubrahmanyan, "Women as Targets in India's Family Planning Policy"; and Phillida Bunkle, "Calling the Shots? The International Politics of Depo-provera," both in *Test-Tube Women: What Future for Motherhood?* ed. Rita Arditti, Renate Duelli-Klein, and Shelly Minden (Boston: Pandora Press, 1984).

26. Bell Hooks puts the issue this way in her *Talking Back: Thinking Feminist, Thinking Black* (Boston: South End Press, 1989).

learn even more about the social and natural orders if we start from the situations of women in devalued and oppressed races, classes, and cultures. Most obviously, we can learn how race, class, and culture shape the situations and visions of women in both dominant and dominated groups. This knowledge is impossible to come by if we start only from the perspective of the lives of women in the dominant groups, for the same reasons that starting from men's lives can not reveal the gendered character of those lives or that thought.

Teresa de Lauretis argues that the female subject is a "site of differences."

> An all-purpose feminist frame of reference does not exist, nor should it ever come prepackaged and ready-made. We need to keep building one, absolutely flexible and readjustable, from women's own experience of difference, our difference from Woman and of the differences among women. . . . However, if I am not mistaken in suggesting . . . that a new conception of the subject is, in fact, emerging from feminist analyses of women's heterogeneous subjectivity and multiple identity, then I would further suggest that the differences among women may be better understood as differences within women. For if it is the case that the female subject is en-gendered across multiple representations of class, race, language, and social relations, it is also the case . . . that gender is a common denominator: the female subject is always constructed and defined in gender, starting from gender. In this sense, therefore, if differences among women are also differences within women, not only does feminism exist despite those differences, but, most important, as we are just now beginning to realize, it cannot continue to exist without them.[27]

To ground claims in women's lives, then, is to ground them in differences "within women" as well as between men and women.

In sum, the logic of the standpoint approaches contains within it both an essentializing tendency and also resources to combat such a tendency. Feminist standpoint theory is not in itself either essentialist or nonessentialist, racist or antiracist, ethnocentric or not. It contains tendencies in each direction; it contains contradictions. And its logic has surprising consequences: the subject/agent of feminist knowledge

---

27. Teresa de Lauretis, "Feminist Studies/Critical Studies: Issues, Terms, and Contexts," in *Feminist Studies/Critical Studies,* ed. Teresa de Lauretis (Bloomington: Indiana University Press, 1986), 14.

is multiple and contradictory, not unitary and "coherent"; the subject/agent of feminist knowledge must also be the subject/agent of every other liberatory knowledge project; and women cannot be the unique generators of feminist knowledge. But exploration of these conclusions must wait for later chapters.

(5) Is feminist standpoint theory excessively modernist? Is it still too humanist and too loyal to Enlightenment beliefs? I have already answered charges from postmodernists that feminist science and epistemology, especially standpoint theories, are damaged by excessive loyalty to humanist and Enlightenment beliefs—through the analysis in Chapter 5 of the diverse grounds claimed for the scientific and epistemological advantage of starting thought from women's lives, through the analysis in Chapter 6 of "strong objectivity," and through the defenses of standpoint theory in this chapter. Pursuing this question further requires closer examination of feminist ambivalence toward the Enlightenment.

The Links between Feminism and Postmodernism

The borders and character of the postmodernist critique of the Enlightenment, its various forms, its relation to modernism (and modernization) are themselves topics of continual debate.[28] Jane Flax has argued that in spite of understandable ambivalence toward the Enlightenment, feminism is—and should recognize that it is—solidly in the terrain of the postmodern. Feminist standpoint epistemology, she says, is one of the theories that are still too firmly and uncritically grounded in faulty Enlightenment assumptions.

> The notion of *a* feminist standpoint that is truer than previous (male) ones seems to rest upon many problematic and unexamined assumptions. These include an optimistic belief that people act rationally in their own interests and that reality has a structure that perfect reason (once perfected) can discover. Both of these assumptions in turn depend upon an uncritical appropriation of . . . Enlightenment ideas. . . . Furthermore, the notion of such a standpoint also assumes that the oppressed are not in fundamental ways damaged by their social experi-

28. One widely cited guide to postmodernisms is Andreas Huyssen, "Mapping the Postmodern," in Nicholson, *Feminism/Postmodernism*.

ence. On the contrary, this position assumes that the oppressed have a privileged (and not just different) relation and ability to comprehend a reality that is "out there" waiting for our representation. It also presupposes gendered social relations in which there is a category of beings who are fundamentally like each other by virtue of their sex—that is, it assumes the otherness men assign to women. Such a standpoint also assumes that women, unlike men, can be free of determination from their own participation in relations of domination such as those rooted in the social relations of race, class, or homophobia.[29]

One can see that Flax, though also sharply critical of nonfeminist postmodernism, has grave doubts about feminist science and epistemology projects.

Other feminist theorists (including some who are attempting to redirect the science traditions) argue that the postmodernist critics offer little help to feminism and that feminists make a big mistake in adopting postmodernist postulates. Nancy Hartsock writes:

> In our efforts to find ways to include the voices of marginal groups, we might expect helpful guidance from those who have argued against totalizing and universalistic theories such as those of the Enlightenment. . . . Despite their apparent congruence with the project I am proposing, these theories, I contend, would hinder rather than help its accomplishment. . . . For those of us who want to understand the world systematically in order to change it, postmodernist theories at their best give little guidance. . . . At their worst, postmodernist theories merely recapitulate the effects of Enlightenment theories—theories that deny marginalized people the right to participate in defining the terms of interaction with people in the mainstream.[30]

Christine Di Stefano, also arguing against the location of feminism in the terrain of the postmodern, finds the great strength of feminist theory and politics in its modernist insistence on the importance of gender.

> Contemporary Western feminism is firmly, if ambivalently, located in the modernist ethos, which made possible the feminist identification and critique of gender. . . . The concept of gender has made it possible for feminists to simultaneously explain and delegitimize the presumed homology between biological and social sex differences. At the same

29. Flax, "Postmodernism and Gender Relations," 56.
30. Hartsock, "Epistemology and Politics," 190–91.

time, however, gender (rather than sex) differences have emerged as highly significant, salient features which do more to divide and distinguish men and women from each other than to make them parts of some larger, complementary, humanistic whole.

She goes on to summarize key aspects of

the feminist case against postmodernism [which] would seem to consist of several related claims. First, that postmodernism expresses the claims and needs of a constituency (white, privileged men of the industrialized West) that has already had an Enlightenment for itself and that is now ready and willing to subject that legacy to critical scrutiny. Secondly, that the objects of postmodernism's various critical and deconstructive efforts have been the creations of a similarly specific and partial constituency (beginning with Socrates, Plato, and Aristotle). Third, that mainstream postmodernist theory (Derrida, Lyotard, Rorty, Foucault) has been remarkably blind and insensitive to questions of gender in its own purportedly politicized rereadings of history, politics, and culture. Finally, that the postmodernist project, if seriously adopted by feminists, would make any semblance of a feminist politics impossible. To the extent that feminist politics is bound up with a specific constituency or subject, namely, women, the postmodernist prohibition against subject-centered inquiry and theory undermines the legitimacy of a broad-based organized movement dedicated to articulating and implementing the goals of such a constituency.[31]

One can begin to clarify some of the issues between feminists by noting the tendency to conflate what I will refer to as "Postmodernism" with "postmodernism." Is the term supposed to refer to the new views arising within and around us as the founding assumptions of the modern West are questioned? Or does it refer to a specific set of claims and practices that have been self- or otherwise identified as Postmodernism?[32] It would simplify matters to say that not all criticisms of or oppositions to Enlightenment philosophy in the present postmodern moment are Postmodernist. Many different social groups are trying to think their way out of the hegemony of modern Western political phi-

31. Di Stefano, "Dilemmas of Difference," 75–76.
32. It is also confusing that philosophers tend to speak of "modernity" as arising during the Renaissance (as in "modern science" and "modern philosophy"), while literary and culture critics apply the term "modernism" to the literary and artistic tendencies of the early twentieth century, and development theorists refer to the "modernization" of different societies at different times in the last 150 years.

losophy and the worlds it has constructed. Feminists of many varieties, groups on the left, movements for Third World peoples in the West and in the Third World, the ecology movement, and the philosophical movement called "Postmodernism" are all critics, albeit in different ways, of Western political philosophy and its sciences. Thus one should not be surprised at the ambivalence of many feminist theorists toward the choice between Enlightenment and Postmodernist projects.

My argument is that such an ambivalence should be much more robust and principled than that identified by Flax and Di Stefano. They attribute to feminists a tentative, hesitant, reluctant ambivalence—one frequently not even articulated—with respect to which side of this dispute feminism should be on. The principled ambivalence I think appropriate would be self-conscious and theoretically articulated, a positive program of belief and action. The rationale for it should not refer primarily to feminist error, or exclusively to intellectual and political inadequacies in the mainstream debate, although that debate *is* woefully inadequate to feminist needs. More important in generating this kind of ambivalence are tensions and contradictions in the worlds in which feminists move. At least some of the tensions between scientific and Postmodernist agendas are desirable; they reflect different, sometimes conflicting, but legitimate political and theoretical needs of women today. The conflict between our different and valuable political projects is just what is creating in feminist thought a necessary ambivalence toward the Enlightenment and toward the beliefs and politics of Postmodernists.

Let me first summarize the possible responses to five charges typically made against feminist science and epistemology by feminist Postmodernists, then suggest that feminist Postmodernism itself contains loyalties (in some respects, excessive) to Enlightenment tendencies.

Postmodernist Tendencies in Feminist Standpoint Epistemology

First, is standpoint theory committed to essentialist assumptions? No, as I argued earlier in this chapter.

Second, must standpoint theory make claims on behalf of a reasoning—a rationality—that is disembodied? On the contrary, it insists that reason is socially located. Our best beliefs as well as our worst ones have social causes, and it is better to begin from some social locations rather than others if one wants to generate less partial and distorted claims about nature and social relations. That is not to say that only persons in female bodies do or can generate such beliefs, let alone that all such persons do or can. But it is to say that one must be engaged in historical struggles—not just a disembodied observer of them—in order to be able to "occupy" such social locations. Those historical struggles make one's arguments "embodied," not transcendental.

Third, must standpoint theorists hold that our "best" representations of the world are transparent to the world—that is, are true? This charge is related to the questions about whether feminist standpoint theory is "too epistemological" and too loyal to science. We can reply here that, on the contrary, we can sort our beliefs into the more versus the less partial and distorted, or into the more versus the less false, without having to commit ourselves to the belief that the results of feminist research are "true." Consider Thomas Kuhn's conclusion that it would be better to conceptualize the history of the growth of scientific knowledge as progress away from falsity rather than toward truth.[33] (Whether this is the best way to think about the history of Western science is a debatable but different point.) Starting research in women's lives leads to socially constructed claims that are less false— less partial and distorted—than are the (also socially constructed) claims that result if one starts from the lives of men in the dominant groups. These "best claims" are themselves socially situated.

Fourth, are the agents of reason and knowledge in the standpoint theory unitary individuals? No, because standpoint arguments claim that knowledge arises from the bifurcated consciousness, the contradictory loyalties, of women who try to fit their understandings of their lives and the understandings they can get through feminist politics into the dominant culture's ways of conceptualizing women's lives. These agents are neither unitary nor conventional "individuals." It is out of the "difference" within each feminist consciousness, as well as within the group "women," that the scientific and epistemological advantage of feminist thought develops.[34]

Finally, does standpoint theory hold that reason and science are progressive? We might best answer this charge by pointing out that there is no "reason" or "science" but only particular, historical forms

33. Kuhn, *Structure of Scientific Revolutions*.
34. Chapter 11 develops further the importance to knowledge-seeking of "bifurcated consciousness."

of reasoning and research traditions, institutions, and practices. Thus, even highly admired modes of thought or research traditions are not all progressive. Thought must always start from some historical social situation and proceed toward some specific historical end. The progressiveness of some forms of reason and science derives from the social purposes and values they advance, not from any values they can embody apart from these historical characteristics. Science and reason contain both progressive and regressive impulses. Historical conditions release or inhibit those impulses.

I do not mean to flatten out or devalue important gaps and conflicts between feminist standpoint epistemology and feminist Postmodernism. I do intend to try to help move the discussion of such gaps and conflicts into yet more fruitful arenas by clarifying where gaps do *not* exist.

The Modernity of Feminist Postmodernism

Feminist standpoint theory is not alone in exhibiting ambivalence toward Enlightenment assumptions. To begin with, how could a feminist theory completely take leave of Enlightenment assumptions and still remain feminist? The critics are right that feminism must at least in part stand on Enlightenment ground. Most obviously, feminist Postmodernists join those they criticize in believing that social progress is desirable and possible and that improved theories about ourselves and the world around us will contribute to that progress. Their own writings, whether or not they overtly dispute these Enlightenment claims, in fact enact them. They debate what those theories should say, whether science and epistemology projects will in fact lead to better conditions for women, and who should get to define what counts as social progress.

Paradoxically, feminist Postmodernism itself may well subscribe to *too many* Enlightenment assumptions. For one thing, in criticizing the very goal of an improved, specifically feminist science and epistemology, it appears to agree with Enlightenment tendencies that all possible science and epistemology—anything deserving these names— must be containable within modern, androcentric, Western, bourgeois forms. But there are good reasons to find this a peculiar assumption. The high cultures of Asia, the Americas, and Africa—those that existed prior to the rise of the North Atlantic cultures—had sophisticated sciences and technologies by the standards of their day (see Chapter 9). The extent of human rationality is neither restricted to nor, most likely, paradigmatically exhibited by the modern West. If other institutions and practices of gaining knowledge have existed outside the modern, bourgeois, androcentric West, why should we assume that no others can exist in the future?

Additionally, the Postmodernist critics of feminist science, like the most positivist of modernity's thinkers, appear to assume that if one gives up the goal of telling one true story about reality, one must also give up trying to tell less false stories. They assume the symmetry between truth and falsity discussed above and in Chapter 6. Feminist thought can aim to produce less partial and distorted representations without having to assert their absolute, complete, universal, or eternal adequacy. Isn't that how we should take the feminist Postmodernists' own analyses? If not, what is the status of their arguments?

Perhaps we should conclude that both feminist science and epistemology proponents and also their feminist Postmodernist critics stand with one foot in the Enlightenment and the other in the present moment—or, rather, the future. That link to the past has problematic and fruitful aspects for both groups. Important differences between them are generated in large part by the different intellectual and political contexts in which they work. They have different histories, partially different audiences, and different goals. Memories or other disputes muddy the psychic grounds on which they meet. At this moment in history, our feminisms need both Enlightenment and Postmodernist agendas—but we don't need the same ones for the same purposes or in the same forms as do white, bourgeois, androcentric, heterosexist Westerners.

I have defended feminist standpoint epistemology against so many critics here, and on so many grounds, that one might think I have been trying to claim it as absolutely perfect. My goal is a more modest one: to show that its flaws are not the ones featured in what we could call the first round of its critical evaluations. The incidence of opposing criticisms of it from groups with conflicting agendas—too objectivist versus too relativist; too epistemological versus too sociological; too critical of science versus not critical enough; too loyal to the Enlightenment versus too Postmodernist—may well indicate that feminist standpoint epistemology really is doing something different and important.

III

"Others"

8

" . . . and Race"?
Toward the Science Question in Global Feminisms

Feminist challenges to conventional beliefs about science, technology, and epistemology have moved from calls for reform to the beginnings of transformative programs for achieving sciences that are both more democratic and more objective. However, most of us who are feminist science critics are of European descent, as are the majority of intellectuals in other fields in the West. We have argued that affirmative action is a scientific and epistemological issue as well as a moral and political one. We have argued that the social group that gets the chance to define the important problematics, concepts, assumptions, and hypotheses in a field will end up leaving its social fingerprints on the picture of the world that emerges from the results of that field's research processes. Since most of us asking the feminist questions have been white and Western, is it not inevitable that we too have left distinctive fingerprints on the images of nature, social relations, science, technology, and epistemology that have resulted from our research?

Several Western feminist critics already have raised questions about the Eurocentrism of the sciences and the social studies of science, and I do not mean to undervalue the significance of their work. In the natural sciences—upon which this chapter primarily focuses—important contributions have been made by Western women of color such as Darlene Clark Hine, by the biracial authorship of Anne Fausto-Sterling and Lydia English, by white women such as Donna Haraway, and by many critics of technologies of reproduction and production.[1] Additionally, science observers who are not primarily identified as feminists have provided important analyses of relations between racist and sexist projects in the sciences.[2] All these writers have challenged the assumption that the history of race relations in the West is irrelevant to accounts of the development and practices of Western sciences and technologies, philosophies and sociologies of science, and epistemologies. European American philosophers have also produced important feminist analyses of racist and Eurocentric assumptions in Western feminist writings, and their arguments are relevant to the science discussions.[3]

After the Fall of Universal Woman

How should a consciousness of race issues inform feminist thought about science? We writers of European descent often say " . . . and race"—meaning, usually, that we *should* also think about the lives of women of color when we talk about "women's situation." We say, "Of course, for women of color, racism creates additional problems." However, we do not ask what happens when we actually do try to add the different problems of women of color to the focuses of existing analyses. Terminological and conceptual problems (usually with political dimensions) arise in attempts to do so. Some are identified below, but a few deserve preliminary comment.

For one thing, should certain U.S. citizens be referred to as "blacks," "Third World people," "people of Third World descent," "people of

1. Darlene Clark Hine, *Black Women in White: Racial Conflict and Cooperation in the Nursing Profession, 1890–1950* (Bloomington: Indiana University Press, 1989); Anne Fausto-Sterling and Lydia English, "Women and Minorities in Science: An Interdisciplinary Course," *Radical Teacher* 30 (1986); Anne Fausto-Sterling, "In Search of Sarah Bartmann" (forthcoming); Donna Haraway, *Primate Visions: Gender, Race, and Nature in the World of Modern Science* (New York: Routledge, 1989). I mention the race of these authors because "Western" does not accurately provide this information.

2. E.g., Sander Gilman, *Difference and Pathology: Stereotypes of Sexuality, Race, and Madness* (Ithaca: Cornell University Press, 1985); Stephen Jay Gould, *The Mismeasure of Man* (New York: Norton, 1981).

3. E.g., Marilyn Frye, "On Being White: Toward a Feminist Understanding of Race and Race Supremacy," in *The Politics of Reality: Essays in Feminist Theory* (Trumansburg, N.Y.: Crossing Press, 1983); Margaret Simons, "Racism and Feminism: A Schism in the Sisterhood," *Feminist Studies* 5:2 (1979); Elizabeth V. Spelman, *Inessential Woman: Problems of Exclusion in Feminist Thought* (Boston: Beacon Press, 1988).

color," "racial minorities," or "African Americans"? There are pros and cons for each term; I use whichever seems most appropriate in the context.

Next, the term "feminist" is even more contentious in discussions of race, gender, and science than it is when race issues are not centered. For many women of Third World descent who are theorists and activists on behalf of women, "feminism" means only the Eurocentric and usually also class-oppressive agendas and practices of economically advantaged women of European descent. For those theorists and activists, "feminism" is part of their problem; it is far too politically and intellectually regressive a term to capture the goals of their own feminisms. Women working on gendered aspects of class issues also sometimes think "feminism" part of their problem. Gay women who have suffered from the homophobic behavior and attitudes of so many heterosexists who called themselves feminists also eschew the term.

In my opinion, progressive feminists who are white, Western, economically advantaged, and/or straight should respect these assessments. We should redirect our analyses of women's situations and our agendas so that they are significantly closer to the more comprehensive ones advocated by women who suffer from more than what some women frequently see as simply "gender oppression." Doing so involves "reinventing ourselves as Other," which is part of the goal of this chapter and book. I continue to speak of the program discussed in this chapter as a feminist one for two reasons. Most readers will not think of "feminism" as too conservative a term in this context. More important, the radically democratic and liberatory tendencies within feminism can be nourished and encouraged, and that is part of my intent.

But in another respect the initial project of this chapter may be regarded as too conservative. When we recollect the limitations of trying to understand women's situation with respect to the sciences, technology, and epistemology by "adding women" to the conventional discourses, shouldn't we recognize that trying to "add women of color" to feminist discussions is similarly problematic? It is—but it also has radical aspects. I think it is worthwhile for both reasons. After all, everyone has to start somewhere when entering a new field; this is certainly more progressive than discussing science and technology without considering the lives of women of color at all. There are important literatures to think about and issues to be raised within the conceptual frameworks of existing feminist discussions. Furthermore, many people are becoming increasingly familiar with the kinds of criticisms of science and technology that (white) feminists have raised. For these people especially, the familiarity of the feminist critiques can prepare the way for an appreciation of the importance of the race issues.

It is not at all easy to "add women of color" to science and technology discussions. The very lack of review essays, books, and courses on this topic makes it difficult to gather the existing literature or to find analyses that reflect on the consequences of these writings for conventional thought. The racism that has silenced and excluded peoples of Third World descent within the sciences and the academic world has made it harder to locate reports of how they have experienced and interacted with Western sciences and technologies and to determine what people of European descent can learn about themselves from the experiences of the others.

Nevertheless, this is a good project also because there are limits to what it can accomplish. The problems we meet should give us a sense of why we cannot *end* with "adding women of color" to the feminist science and technology discussions, even if that is the necessary place to start. The effort can help move our thought toward more ambitious understandings. From their beginnings, the "add women and stir" approaches have tended to burst through their predesigned boundaries and raise usefully uncomfortable questions about the canons and bodies of knowledge to which they were supposed to be merely adding information. The "add women of color" efforts can have this effect as well. So while there are more radical ways to begin to take a global feminist perspective on Western science and technology, there are good reasons to recognize this approach as an appropriate and valuable one that will lead, I hope, to the radical questions.

The next section shows some ways to go about adding issues about women of color's experiences of and interactions with Western science and technology to the topics that have appeared most important in the existing critiques by (primarily) white, Western feminists. Then I turn to more general issues of what we miss when we settle only for adding the lives of women of color to existing analyses. The ways in which such a framework is inadequate to address their lives are also ways in which it yields inadequate accounts of *everyone else's* relationships to issues of science and technology and gender. Any analysis that provides only partial and distorted accounts of the lives of women of color can provide only partial and distorted accounts of the lives of white men, men of color, and white women as well. In short, when feminists deconstructed universal man, they also undermined the possibility of speaking about or from the perspective of his faithful companion, universal woman. The nonexistence of universal women has implications for critiques of science and technology.

From the Women of Color Question in Science to the Science Question in Global Feminisms

What can be learned by trying to add women of color to the existing concerns of the feminist science analyses? First, one cannot assume that there is a single story to be told about these women's experiences of and interactions with sciences and technologies. Even in the United States the effects of individual and institutional racism have been different for women in different races, classes, and cultural groups. We cannot generalize about women of color in the United States, and we cannot generalize about the even more diverse group of women in the Third World. There is no typical woman of Third World descent to follow in her experiences as producer, consumer, or object of study of Western sciences and technologies. The focus here is primarily on American women of African descent, with such glimpses of the lives of other women of Third World descent as the literature offers.[4]

Women of Third World Descent in the Sciences

In the United States the opportunities in the sciences for women of color have been even more vigorously restricted than for white women. Specific information is difficult to come by. It is frequently absent from accounts of the experiences of "women," which tend to focus on white

4. Good bibliographies can be found in Anne Fausto-Sterling, "Women and Minorities in Science: Course Materials Guide," 1987 (available from Fausto-Sterling, Division of Biology and Medicine, Brown University, Providence, R.I. 02912); Nancie L. Gonzalez, "Professional Women in Developing Nations: The United States and the Third World Compared," in *Women in Scientific and Engineering Professions*, ed. Violet B. Haas and Carolyn C. Perrucci (Ann Arbor: University of Michigan Press, 1984); the sections on "Women of Color" and "Women in the Third World" in Ruth Hubbard, M. S. Henifin, and Barbara Fried, eds., *Biological Woman: The Convenient Myth* (Cambridge, Mass.: Schenkman, 1982).

women; it is absent from the discussions of "minorities," which usually focus on men's lives. Quantitative data about women of color is often aggregated within the data about "women" or "minorities," or sometimes even "women and minorities." Nevertheless, some intriguing glimpses of African American women's lives in the sciences and technology are available.

A very few black women have managed to gain the education and credentialing for careers in the natural sciences, and these have been disproportionately in medicine. Darlene Clark Hine has written about the 115 black women who received M.D. degrees in the United States in the quarter-century after the end of slavery, degrees granted by the New England Female Medical College in Boston, the Women's Medical College of Philadelphia, the New York Medical College for Women, and others. Hine points out that these women led lives different from those of other physicians—men or white women—in several ways. Most obviously, "they were an integral part of the black communities in which they practiced," working at black colleges and in community clinics and hospitals, becoming clubwomen, and successfully combining their lives as wives and mothers with their careers as physicians—thereby challenging the prevailing belief that higher education and professional training made a woman less feminine (though femininity clearly had a different meaning in African American communities than it did among whites, as Sojourner Truth pointed out). Moreover, they also founded an array of health-care institutions: "They established hospitals and clinics, trained nurses, taught elementary health rules to students and patients, and founded homes and service agencies for poor women and unwed girls of both races." Though this small group of professionals came primarily from the upper echelons of black society, each one used her education and skills for the benefit of black people: "These early black professional women played an undeniably significant role in the overall survival struggle of all black people."[5]

By the 1920s there was a notable reduction in the number of black women physicians. Hine suggests two reasons: the convergence of renewed forces of racism, sexism, and professionalization no doubt was

5. Darlene Clark Hine, "Co-Laborers in the Work of the Lord: Nineteenth Century Black Women Physicians," in *"Send Us a Lady Physician": Women Doctors in America, 1835–1920*, ed. Ruth J. Abram (New York: Norton, 1985), 117.

responsible in part; moreover, "instead of entering the medical profession, the aspiring, career-oriented black women began focusing on nursing as a more viable alternative for a professionally rewarding place in the American health care system."[6] From other sources, it becomes clear that foundations put pressure on educational institutions to limit the career aspirations of blacks in the sciences and medicine.[7] Hine's recent study of black nurses from 1890 to 1950 shows the immense struggles it took for black women to gain access to nursing education, just as black physicians were being denied access to staff positions in white hospitals and black patients were being denied access to medical treatment. Because of this discrimination and exclusion, black communities developed a parallel system of training schools and hospitals.[8]

When the topic is American scientists of African descent, the discussion of "why so few" women or men requires a more comprehensive context than when scientists of European descent are the primary population under consideration. It is amazing that any African Americans achieved scientific careers in fields other than medicine prior to World War II when one considers how severely limited graduate science education was for them.[9] Indeed, even public high school education was unavailable for some urban African Americans until as late as the mid-1960s. In the early 1960s, some cities and states in the South still provided no public education at all for blacks past the eighth grade. A recent lecturer on my campus described the Tallahassee, Florida, "Negro High School" that he had attended: segregated elementary schools asked African American parents to permit their children to be "left back" in eighth grade for four years so that their teachers could provide them with the knowledge and skills equivalent to those that European American citizens of Tallahassee could get from the four-

6. Ibid.

7. E.g., Kenneth Manning, *Black Apollo of Science: The Life of Edward Everett Just* (New York: Oxford University Press, 1983).

8. Hine, *Black Women in White.*

9. Accounts of these past and present struggles are becoming available. See, e.g., Sara Lawrence Lightfoot, *Balm in Gilead: Journey of a Healer* (New York: Addison-Wesley, 1989); Manning, *Black Apollo;* Aimee Sands, "Never Meant to Survive: A Black Woman's Journey" (interview with Evelynn Hammonds), *Radical Teacher* 30 (1986); Ivan Van Sertima, ed., *Blacks in Science: Ancient and Modern* (New Brunswick, N.J.: Transaction Books, 1986); and others listed in Fausto-Sterling, "Women and Minorities: Course Materials Guide."

year high school curriculum. Then they encouraged at least some of these eighth grade "graduates" to apply to the most rigorous colleges. The speaker, a college dean, had undergraduate and graduate degrees from Harvard, which he had entered straight from his segregated eighth grade in the late 1950s.

The situation in engineering and technological innovation, which have always been closely related to scientific practice, was almost as dismal. Free African Americans were legally able to patent the important technological innovations they made from the time of the first U.S. Patent Act (1790), but few received patents before the Civil War. Slaves, women as well as men, devised many inventions, but until the end of the Civil War, neither slaves nor their owners—who frequently tried— were permitted to receive patents for slaves' inventions.

Other factors also contributed to the scarcity of scientists of African descent in the United States. In both the natural and social sciences, if one divided the whole history of Western scientific projects into those that appear to have little to do with improving African American lives, those that have the consequence (intended or not) of deteriorating the quality of those lives, and all the rest, one would find very little scientific work in the first and third categories. The sciences have not in general had good effects on African American lives.[10] Why should African Americans want their young people to enter such fields as physics, chemistry, biology, mathematics and engineering in light of the resources that the sciences have provided for racist policies and practices? Until recently, there has been no push in educational institutions at any level—either primarily black or integrated ones—to encourage African American youth to pursue science careers. African American communities have insisted that their children be educated in ways that had direct benefits to African Americans.[11]

Is it fair for African Americans today to have to think that they should enter science and technology only if doing so will improve the conditions of black lives? After all, one might think, European Americans are not expected to calculate whether their work will benefit white communities before they are permitted to feel good about enter-

10. For the social sciences, see, e.g., Robert V. Guthrie, *Even the Rat Was White: A Historical View of Psychology* (New York: Harper & Row, 1976); Joyce Ladner, ed., *The Death of White Sociology* (New York: Random House, 1973).

11. I am indebted to Evelynn Hammonds for emphasizing the importance of this issue.

ing physics, chemistry, or engineering. The community service ethic has often appeared to be an unjust burden, especially when European Americans use this kind of argument to direct talented African Americans out of predominantly white institutions and fields and into black ones. "You have a duty to your people," white educators and foundations said to the "Black Apollo of Science," Ernest Everett Just. "Go teach undergraduates at Howard University, instead of pursuing a research career in physics."[12]

It is no more appropriate for me, a European American, than it was for those to whom Just had to listen to pronounce on which careers African Americans should pursue. Nevertheless, we can note that while the argument against requiring a community service ethic for African Americans can appear compelling, the situation is not quite as parallel to the situation for European Americans as the argument implies. Not only have Western science and technology been complicitious with overt racist social agendas, but they also encode racial messages in the very definition of their most abstract projects. European Americans too enter science as a "calling," inspired by the heroic tales of the Great Men whose accomplishments supposedly demonstrate the highest intellectual achievements of the "human" race, whose work has purportedly brought such great benefits to "humankind." Among these benefits, they are told, are powerful rational defenses against irrational superstition, "primitive" ways of thought, and oppressive and exploitative politics that are so often supported by "the masses." Decoding the racial messages (as well as the gender and class ones) in these conventional justifications for entering the arduous training that careers in the sciences require reveals that European Americans are indeed enticed to enter science in order to improve conditions for *their* race. Not only is a "white service ethic" apparently supposed to be part of the implicit appeal of a career in science for European Americans, but—unlike the black service ethic—it contains not-so-hidden racist messages.

In spite of the Western sciences' racist agendas and coding of their mission, I think we can see why it is important that there be more African American women and men in the sciences. All the arguments made in Chapter 3 for increasing the numbers of women in the sciences are also arguments for increasing the numbers of African Americans

12. See Manning, *Black Apollo*.

and other persons of color. Social justice mandates making available to African Americans too the social benefits that come from careers in the sciences. Moreover, scientific literacy is crucial for anyone to function effectively in society today. Furthermore, African American young people need the mentoring and role modeling in the sciences that all scientists should provide for them but—given the racial stratification of U.S. society and the negative messages most European Americans send out about African Americans' intellectual abilities—cannot be depended upon to provide. African Americans and other people of color should also have access to the resources for designing sciences and technologies that will benefit their communities. And the sciences stand a far better chance of losing their racist agendas and codes if there is a significant African American presence among scientists. Last but not least, it is hard to imagine advancing democratic tendencies in Western societies without the prominent participation of African Americans and other persons of color—women and men—in directing scientific and technological policy.

This is just one area where it is important to look carefully at the differences in the situations of women of color in different societies. In a few Third World countries (for example, in South America and the Caribbean), just as in the West, social stratification has sometimes permitted women in the upper classes to gain education and careers little different from those of their brothers.[13] Consequently, the women already in the sciences in these countries are frequently members of an economic and political elite that is far less likely to challenge the West's ideas about how the sciences and technologies should function than would less economically and politically advantaged members of those societies. Nevertheless, the new histories show again and again that women in privileged classes have frequently been a powerful force for progressive social change, contrary to the conventional opinion of some men on the left that they are always a conservative force in history. Even in the dominant groups, women's relatively marginal status as women, coupled with the ethic of caring for others that is expected of these women too, often makes possible their sympathetic identification with the needs and interests of people more marginal than they. Many activists who have worked to improve the conditions

---

13. Gonzalez, "Professional Women in Developing Nations."

of racially and economically marginalized women in the United States and Europe have come from the upper classes.

How are the cautions raised in the conventional feminist literatures about women's experiences in the social structure of science pertinent here? First, focusing on a few African American women "worthies" who have managed to gain access to European American science institutions does not tell us much about the vast majority of such women who work in the sciences or may have aspired to scientific and technological careers. Even more than the lives of Great European American Men and the few Great European American Women who have achieved recognition in the sciences, these lives are by definition extraordinary. Furthermore, the histories of women in the sciences have tended to focus disproportionately on elites, as did their "prefeminist" counterparts. This has the effect of making almost invisible any presence of women of color in the sciences. A search for Great African American Women in the sciences would suffer from the fact that it tried to establish the individuality of a few women at the expense of the masses whose efforts made those achievements possible. Such an approach would advance Eurocentric preoccupations with individualism and meritocracy at the expense of understanding and supporting the collective and community-focused ways in which science and technology have in fact been practiced in our own and other cultures.

Another challenge is to resist the temptation to assess the contributions of African American women to science and technology only from the perspective of what European American elite men or women count as scientifically and technologically interesting and valuable. Hine's studies of black women physicians' educational and social work to benefit their communities' activities and of black women's later focus on nursing rather than doctoring highlight the importance of kinds of scientific activities that are devalued in dominant white, Western men's science circles. It is also important to discover what their achievements and participation in science and technology mean to women of African descent.

I have been discussing primarily African American *women*, but it sounds odd to talk of African American women as if their situation has not been fundamentally shaped by conditions they have shared with the men of their communities. The discussions of the participation by women of European descent in the history of science and technology gained their point primarily through contrasts between the opportunities and experiences of such women and those of the men of their race, class, and culture. But even though comparison of the situation of women of African descent with their brothers' situations can be illuminating, focusing only or even primarily on that gender contrast distorts the account. In order to understand the participation of African American women in the sciences and technologies, it is necessary to set that story within the general patterns of African American history—not just of European American history. Black women are not "like white women, only different"; they are created through and creators of a broader history of race relations. The histories of women in the sciences tend to be embedded largely in European American history, which is understandable if one's focus is on what the dominant culture is willing to count as science. But if science is part of the larger society—if society is in science, as I argued in Chapter 4—we need to understand how patterns of race relations in the larger society have had consequences for African Americans' experiences of and interactions with the sciences.

Finally, we can wonder whether the entrance of large numbers of African American women and men into positions of authority in the sciences and their technologies would leave the content of scientific claims and the logic of research and explanation unchanged. I argued in earlier chapters that gender equity is a scientific and epistemological issue as well as a moral and political one. Is this not also true for racial equity?

The Effects of Science and Technology on Women of Color

In the United States, African American women suffer from more extensive and intense forms of the same bad effects of Western science and technology that European American women experience (see Chapter 2). But there are also differences in quality as well as quantity between these life patterns. For one thing, racism and the higher rates of poverty it ensures have deprived many African American women of some of the resources that European American women can call upon in order to negotiate with scientific, technological, and medical establishments.

One place to see these patterns is in reproductive policies and practices. In the United States a disproportionate number of African American women are among those who have been sterilized without informed consent. The same racism and poverty that are responsible for the high numbers of sterilizations also make it more difficult for African American women to gain access to safe abortions. The Supreme Court has determined that Medicaid, the federal health-care program for poor people, may not finance abortions. Clearly, what European Americans fundamentally cannot tolerate is that African American women and men should make their own reproductive choices. This is not surprising if one reflects on a fact so obvious that it is hard to get into focus: in a race-stratified society, reproduction in both its social and biological aspects (and the biological are themselves shaped by the social) reproduces races, not just race-anonymous infants. Consequently, it *would* be surprising—a fact requiring scientific explanation—if reproductive policies in a racist society did *not* contribute to determining who will live and who will not. And "reproductive policies" involve not only contraception and birthing but also the distribution of opportunities for good health at later stages in people's lives.[14]

Do European American women benefit from racist reproductive policies, technologies, and practices? It is not pleasant for European Americans—let alone feminists in this group—to contemplate the question, but to whatever extent the allocation of social and material opportunities and resources is a zero-sum situation, the answer would appear to be affirmative.[15] More European American than African American females survive in good health as a consequence of these kinds of policies. Moreover, the relative resources available to European American, economically advantaged women and to poor women of Third World descent affect their lives as reproducers, not just as the "reproduced." When reproductive technologies are in the wrong hands, European American women's benefits can occur at direct cost to poor women and women of color. Some have feared that the Baby M. case could be the forerunner of the use of poor and Third World women's wombs to produce children for economically advantaged European American couples. As another example, contraceptives not yet regarded as safe for U.S. mainland use were first tested in Puerto Rico,

14. One good recent account is "Currents of Health Policy: Impacts on Black Americans," *Milbank Quarterly* 65 (1987), spec. supp. 1–2.

15. Phyllis Palmer, "White Women/Black Women: The Dualism of Female Identity and Experience," *Feminist Studies* 9:1 (1983), is one influential discussion of the relationship between African American and European American women's lives today.

giving mainland women the ability to increase their control of reproduction at the cost of increased risks to the health of many Puerto
Rican women.

This kind of phenomenon is not new. In the nineteenth century,
African American slave women were used by at least one prominent
physician and medical researcher, J. Marion Sims, for experimental
gynecological surgery; moreover, he did not use the anesthetic available at the time in his hundreds of experiments. Two of his "patients,"
one a slave and the other a poor Irish woman, actually survived some
thirty such operations.[16] Since it has been a widespread custom in
medical research for the poor to be used in scientific and medical
experiments, it should not be surprising that practices that would not
at the time have been permitted on poor people of European American
descent were conducted on women of color.

Should this kind of scientific work, work that is without question
undertaken in order to advance the growth of knowledge, be regarded
as experimentation or as torture? Is there any standard within science
that will enable us to tell the difference between the two? What are the
obligations of the scientist to report or to take a moral stance on the
results of research gained in such a manner? These are the terms in
which critics are beginning to discuss the many routine references in
recent medical literature to the reports by "Nazi doctors" of the results
of their experiments on Jewish prisoners in the concentration camps—
experiments performed to improve the life opportunities of German
citizens.[17] For example, prisoners were submerged in icy water and the
characteristics of their approaches to death were observed in order to
gain information that might enable German sailors and flyers to survive in northern seas. That these experiments should not have been
conducted goes without saying and thus is not the issue raised in recent
discussions. Nor is the issue whether to use the information, since
knowing the "facts" produced through this "research" can help save
lives. The issue is whether and how the useful information thus produced should be cited in subsequent scientific reports. To repeat the
point, is it the result of experimentation or torture? And how can we—
or science—tell the difference between the two? Are these not also the

16. G. J. Barker-Benfield, *The Horrors of the Half-Known Life* (New York: Harper &
Row, 1977), 96, 102.
17. See "Commentary by Naomi Scheman" (on Sandra Harding, "The Method Question"), *American Philosophical Association Newsletter on Feminism and Philosophy*
88:3 (1989).

terms in which discussions of reproductive experimentation on women of Third World descent should be framed?

If "reproductive policies" are construed more generally as those policies that determine who will live and who will die throughout the life span, many differences in the resources available to European Americans and African Americans become visible. In the United States the 1980s widened the gap between the races in the kinds of reproductive policies the government has supported. As involuntary sterilization and the withdrawal of social services was escalating for African American and poor women, there was a decrease in support for the kinds of regulations and services that would permit European American and middle-class women to engage in full-time wage labor outside the household. The government has retreated in its support of affirmative action and equal opportunity policies and simultaneously decreased its support of social services for children, the aged, and the sick. This has the effect of moving back into the household on a full-time basis the care of children, the sick, and the old for whom the availability of public services had begun to permit women to work for wages. The women who are discriminated against in the job market now have plenty of unpaid but socially necessary labor to keep them busy at home. This appears directly as a class issue but has more bad consequences for African American than for European American households.[18]

Another focus for criticism has been on dangerous, inappropriate, and exploitative technology transferred to the Third World with particular consequences for women. Infant formulas were marketed in Africa without sufficient instruction in how they had to be used—with the consequence that the health of African babies was endangered. Depo-provera and intrauterine devices have similarly been sold to Third World governments—even forced upon them as a condition of continued financial aid from the West—with little concern for educating Third World women and men in their use and even less concern for the general health of the women using them.[19] Residual supplies of

18. See Gisela Boch, "Racism and Sexism in Nazi Germany: Motherhood, Compulsory Sterilization, and the State," *Signs* 8:3 (1983), for a similar scenario in Nazi Germany.

19. See esp. the essays in Rita Arditti, Renate Duelli-Klein, and Shelly Minden, eds., *Test-Tube Women: What Future for Motherhood?* (Boston: Pandora Press, 1984); and in Patricia Spallone and Deborah Steinberg, eds., *Made to Order* (New York: Pergamon Press, 1987).

drugs declared unsafe and consequently prohibited from sale in the West have been dumped in Third World markets.

The literature on women and development is full of examples of technology transfers that are economically and politically advantageous to the West but further deteriorate the material and social resources available to women (and men) in the receiving cultures.[20] When farming technologies and education are provided for men in cultures where farming has traditionally been done by the women, agricultural production declines, because Third World men are no more enthusiastic than their Western brothers about the prospect of spending their days doing "women's work." "Modern" water supplies and cooking technologies are introduced that have the consequence of making more work for women. The so-called "green revolution" replaces the production of crops for local consumption with cash crops for export, creating economic and subsequent nutrition problems for the indigenous population—especially for women, who are often both the food providers and the last in the family to eat. It is increasingly clear that the costs of so-called "development" in the Third World are disproportionately born by poor women.

In the First World as well, science's assistance in "rationalizing" labor processes has succeeded in deskilling labor in many occupations, with the consequence that those who are least powerful in the labor market are moved into whatever work is most degraded and least rewarded. The low wages African Americans can be paid make it possible for bosses to get away with cheaper, more hazardous, more labor-intensive technologies than would otherwise be required. They can also get away with providing no health insurance, no maternity leaves, and no retirement plans. "Runaway factories" have been moving textile, electronics, and other manufacturing industries into the Sunbelt, where African American, Hispanic, and some European American workers are hungrier, unions are weaker, and state governments promise to keep labor regulation low. The increase in the inter-

20. Good sources are Susan Bourque and Kay B. Warren, "Technology, Gender, and Development," *Daedalus* 116:4 (1987); ISIS Women's International Information and Communication Service, *Women in Development: A Resource Guide for Organization and Action* (Philadelphia: New Society Publishers, 1984); Maria Mies, *Patriarchy and Accumulation on a World Scale: Women in the International Division of Labour* (Atlantic Highlands, N.J.: Zed Books, 1986); Maria Mies, Veronika Vennholdt-Thomsen, and Claudia von Werlhof, *Women: The Last Colony* (Atlantic Highlands, N.J.: Zed Books, 1988).

national division of labor has meant that "offshore industries" controlled from the West are located in Third World countries, where women—who are exploited by their families as well as by the government and by employers—can work in textile or electronic manufacturing or in computer processing "piecework" to produce products and services for Western markets far more cheaply.[21] I will not even begin to review the literature detailing these practices but only argue that it is important to one's thinking about the impact of Western sciences and technologies on the lives of women of Third World descent. Certainly, such science has not been used "for the people," in Galileo's phrase— at least not if "the people" means all of the world's peoples.

Accounts of how scientific technologies have affected the lives of women of Third World descent must be situated within an understanding of Third World history more generally. Accounts showing the effects of Western sciences and technologies on Third World peoples should recognize that they have often affected men and women in different ways. All the cautions I raised above against trying to extract the lives of women of Third World descent from their historical contexts in order to "add" them to accounts of First World women's lives apply here as well.

In Chapter 4 I argued against the temptation to think that a core of "pure science" can be extracted from its Eurocentric and racist institutionalization in the modern West. One might still think that only beginning with the massive infusion of public monies into the hard sciences and their technologies since World War II has the "pure" core of science begun to appear so shrunken compared with the expanding array of "impure" misuses and abuses. One might suppose that the gynecological experimentation on African American slave women, which occurred long before Sputnik, was simply the exception that proves the rule; most scientists were not engaged in such blatantly racist practices. No doubt this is true. Nevertheless, certain kinds of race relations have linked modern Western science from its beginnings with the advance of imperialism and colonization in the Third World (see Chapter 9).

Whether or not individual European Americans in fact have *intended* to support the exploitation of people of Third World descent

---

21. In addition to the sources cited in note 20, see Barbara Ehrenreich and Annette Fuentes, "Life on the Global Assembly Line," *Ms.*, January 1981.

through their scientific beliefs and practices (and there is no doubt that many have), science as an institution has functioned in ways that are exploitative. The recognition is growing that the resistance to acknowledging or critically examining the origins, consequences, values, and interests of scientific projects is part of the irresponsibility of positivist tendencies in the sciences and science studies—tendencies (discussed in earlier chapters) that have been useful to antidemocratic tendencies in the state and the economy. Even the National Academy of Sciences is now concerned to make sure everyone understands that "human values cannot be eliminated from science, and they can subtly influence scientific investigations," and that "science and technology have become such integral parts of society that scientists can no longer abstract themselves from societal concerns."[22]

Bias against Women of Color in Research Results

Historians of science and biologists have pointed out that the very same arguments and research projects have been used to disseminate both sexist and racist stories about human evolution. The current uses of sociobiology for racist and sexist purposes is but the latest of these attempts to restore the racial and gender status quo of the 1950s and earlier against the threats posed by rising feminist and Third World movements.[23] And evolutionary theory is just one area of biology in which the combination of racist and sexist assumptions has distorted the image of women of Third World descent.

In the social sciences, several decades of research have produced analyses criticizing and providing alternatives to the partial and distorted views of women of color characteristic both of conventional and European American feminist social theory and research.[24] These anal-

22. National Academy of Sciences, *On Being a Scientist* (Washington, D.C.: National Academy of Sciences Press, 1989), 6, 20.

23. See, e.g., Ruth Bleier, *Science and Gender* (New York: Pergamon Press, 1984); R. C. Lewontin, Steven Rose, and Leon J. Kamin, *Not in Our Genes: Biology, Ideology, and Human Nature* (New York: Pantheon Books, 1984).

24. A small sample of these includes Patricia Hill Collins, "Learning from the Outsider Within: The Sociological Significance of Black Feminist Thought," *Social Problems* 33 (1986); Patricia Hill Collins, "The Social Construction of Black Feminist Thought," *Signs* 14:4 (1989); Angela Davis, *Women, Race, and Class* (New York: Random House, 1981); Bonnie Thornton Dill, "Race, Class, and Gender: Prospects for an All-Inclusive Sisterhood," *Feminist Studies* 9 (1983); Paula Giddings, *When and Where I Enter: The*

yses have the effect of challenging cherished assumptions in the sciences and the philosophy and social studies of the sciences. They challenge the devaluation of the "context of discovery" for shaping the results of research; they challenge the ability of scientific method as it is generally defined to identify and eliminate distorting biases; they challenge the narrow and misleading ways in which objectivity is conceptualized; they challenge received views about the scientific consequences of equity and affirmative action programs; they challenge the adequacy of attempts only to "add women" to the existing subject matters rather than also to transform their conceptual frameworks.

It needs to be stressed that a new understanding of European American women and men can also be found in this literature. European American women have a race too, whether or not we recognize or announce it. We tend to "speak it," to "act it out"—intentionally or not—if we do not learn how to think from the lives of women of Third World descent as well as our own in analyzing "women's situation." This unexamined racial identity not only leads to distorting images of Third World peoples; it also leads to partial and distorted images of women and men of European descent.

Racial and Sexual Meanings of Science and Nature

A few writers have begun to look at science's participation in attributing simultaneously sexualized and racialized meanings to nature and inquiry. Sander Gilman has examined the ways racism and sexism used each other as resources in constructing images of women and "other races" that legitimated their exploitation and domination by men of European descent. For example, he explains how Western medical writers developed a network of associations about the sexual and moral deviance and abnormality of black South Africans, southern European women, prostitutes, and lesbians. In this way they were able to gain widespread support for the idea that womanhood itself was a source of disease and moral corruption. Nancy Stepan has shown that racial and sexual stereotypes have been conjoined in the sciences to the detriment of women and Third World peoples. Anne Fausto-Sterling

Impact of Black Women on Race and Sex in America (New York: Bantam Books, 1985); Maxine Baca Zinn, Lynn Weber Cannon, Elizabeth Higginbotham, and Bonnie Thornton Dill, "The Costs of Exclusionary Practices in Women's Studies," *Signs* 11:2 (1986).

has explored further the meanings that the woman derogatorily named the "Hottentot Venus" had for nineteenth-century Europe.[25]

Donna Haraway has looked at the racist sexism and sexist racism characteristic of primatology, which has provided important scientific legitimation for the association of African women with wild animals in contrast to European and European American women's association with civilization. Haraway identifies the very different preoccupations of Indian, Japanese, and American primatology research and suggests that they can be accounted for by cultural differences in the conceptualization of nature and society. Race and gender meanings are clearly but complexly part of these contrasting conceptions.[26] Literary critics too have identified the conjoined racist and sexist agendas to be found in the history of travel reports and other colonial discourses about exotic nature and exotic peoples.[27]

Chapter 2 noted that metaphorical meanings accumulate social legitimacy—moral resources—for the theories they permeate.[28] They constitute not merely rhetorical decorations or heuristic devices but serve to direct research processes in certain directions—as did the metaphor of nature as a machine—and help to select what will be counted as legitimate evidence for hypotheses. Perhaps the identification of racial themes in scientific literatures should lead to a reexamination of the metaphors familiarly associated with gender in order to see whether these also have racial meanings. For example, how does racism shape the conception, in Bacon's and Machiavelli's figures of speech, of the "wild women" who will threaten man's fate unless controlled by men? Is the apparent wildness of European women enhanced by the association—which can be found especially in writings about women of color—of women with a sexuality that appears to men as "the wild" ("the dark continent," in Freud's words)?[29]

25. Gilman, *Difference and Pathology;* Nancy Stepan, "Race and Gender: The Role of Analogy in Science," *Isis* 77 (1986); Fausto-Sterling, "In Search of Sarah Bartmann."

26. Haraway, *Primate Visions,* esp. chap. 10.

27. See, e.g., Henry Louis Gates, Jr., ed., *"Race," Writing, and Difference* (Chicago: University of Chicago Press, 1986); Edward Said, *Orientalism* (New York: Random House, 1978); V. Y. Mudimbe, *The Invention of Africa: Gnosis, Philosophy, and the Order of Knowledge* (Bloomington: Indiana University Press, 1988).

28. See the discussions of how metaphors provide social and scientific resources in Sandra Harding, *The Science Question in Feminism* (Ithaca: Cornell University Press, 1986).

29. See Sharon Tiffany and Kathleen J. Adams, *The Wild Woman: An Inquiry into the Anthropology of an Idea* (Cambridge, Mass.: Schenkman, 1985).

Eurocentric Epistemologies

Less partial and distorted descriptions and explanations of nature and social relations tend to result when research starts from the lives of women of Third World descent rather than only from the lives of men or of women of European descent. In a society and a world stratified by race, class, and gender, it is impossible for the sciences to maximize objectivity when the scientific community shares elite social interests and values. The outpouring of recent critiques of the Eurocentric political, social, psychic, and economic projects of the natural and social sciences makes clear how partial and distorted have been the images of nature and social relations produced by what were once thought to be the most rigorous applications of scientific method. These criticisms and alternative accounts demonstrate that equity issues are not only moral and political, as usually perceived, but scientific and epistemological as well.

In feminist analyses, with few exceptions, it was not until women of Third World descent had access to publication, university positions, and disciplinary conferences that the perspective from their lives on Western life and thought could be heard by Western feminists. Here the kinds of questions that shaped the construction of feminist empiricism and the feminist standpoint theories are pertinent. Are the criticisms of science and technology from the perspectives of the lives of women of Third World descent intended to extract a "bad science" from its shell of less than fully rigorous practices? Or is there something more fundamentally wrong with European American scientific theories of knowledge? Do we need an African American or Third World or global feminist standpoint epistemology to account adequately for the racism of science and technology and for the greater power of the feminist accounts that are constructed from the perspective of the lives of women of color? Just what is it about these women's lives that gives the epistemological advantage?

Drawing on the discussion in Chapter 5, one can argue for the scientific and epistemological advantage of starting from the lives of those who have been devalued, neglected, excluded from the center of the social order; who generate less interest in ignorance about how the social order works; who provide perspectives from the other side of racial struggles; who enable a different perspective, one from everyday life; who in some cases provide "outsider within" perspectives; who mediate relations between nature and culture in ways different from those of European American women; and whose activities provide particularly illuminating understandings at this moment in history.

Exploration in more detail of exactly how starting from the lives of women of Third World descent can maximize objectivity as it decreases the partiality and distortion of scientific and scholarly images of nature and social relations is a project already well under way in the writings of historians, sociologists, literary critics, economists, political scientists, and others. My point here is to argue for the necessity for *all* research and scholarship to learn how to ground thought in this way. In particular, feminist research will continue to produce unnecessarily partial and distorted accounts if it does not take seriously the standpoint injunction to start from women's lives—and to understand that to mean from *all* women's lives. Attention to institutionalized and structural relationships between the lives of women in different races, classes, sexualities, and societies is crucial for reducing partiality and distortion in accounts of nature and social relations.

Chapter 11 explores further just what it means for a woman of European descent to start her thought in the lives of women of Third World descent. Meanwhile, there are more general issues to be raised about the advantages that accrue if we begin to analyze Western sciences and their technologies from the perspective of the lives of women of color.

Integrating Race, Class, and Gender

As suspected, the additive approaches to race issues could no more be contained within the terrains one might have envisioned for them at the start than could the "add women and stir" approaches to gender issues. Terms have slipped away from their familiar referents: it turned out that in order to discuss science and technology, one had to consider the organization of wage labor and patterns of "de-development" in the Third World. One could not focus only on individuals' prejudices against people of color but had to consider also the agendas and practices of institutions. One could not focus only on women of Third world descent and their lives but had to look also at the lives of Third World men and of European Americans—women as well as men.

New questions about the conventional feminist approaches to science and technology have arisen: do Western women benefit from the misuses and abuses of the sciences and their technologies in the lives of women of Third World descent? Are some of the gender metaphors in Western science discourses also racial metaphors? What conceptual shifts, with what consequences, must occur when one moves past the additive approaches in order to integrate gender, race, and class analyses? What does it mean to try to provide only integrated analyses?

We can begin by noting what such an attempt is not. Maxine Baca Zinn and her colleagues have identified three common and problematic approaches to race and class issues in the writings of white, middle-class, feminist social science researchers. Since this book is concerned with what may be called the "social sciences of the sciences, technology, and the production of knowledge," such criticisms are pertinent. One kind of analysis, say these critics,

> treats race and class as secondary features in social organization with primacy given to universal female subordination. Such thinking establishes what is taken to be a common feminist ground and labels any divergence from it, in Phyllis Palmer's phrase, a "diversionary special interest."

It is simply not true that gender relations create a set of human experiences that are more important than those created by such other inequalities as race and class, but this is what is implied if we make gender relations primary. A second problematic approach

> acknowledges that inequalities of race, class and gender generate different experiences and that women have a race-specific and a class-specific relation to the sex-gender system. However, it then sets race and class inequalities aside on the grounds that, while they are important, we lack information that would allow us to incorporate them in the analysis. As Bonnie Thornton Dill puts it, inequalities other than sex and gender are recognized, but they are not explicated. After a perfunctory acknowledgment of differences, those taking this position make no further attempt to incorporate the insights generated by critical scholarship on race and class into a framework that would deal with women generally.

The third approach

> focuses on descriptive aspects of the ways of life, values, customs, and problems of women in subordinate race and class categories. Here differences are detailed with little attempt to explain their source or their broader meaning. Such discussions of women are [according to

Margaret Simons] "confined to a pretheoretical presentation of concrete problems."

If feminists wish to integrate race, class, and gender issues, they must transform our largely separate theories of the origins and natures of gender, race, and class hierarchy into a single theory. Baca Zinn and her coauthors recommend that "an approach to the study of women in culture and society should begin at the level of social organization. From this vantage point one can appreciate the complex web of hierarchical social arrangements that generate different experiences for women."[30]

In science and technology, that means beginning with the social organization of science and technology institutions, including how they are economically and politically located in race and class as well as gender relations. The ways in which European American people in the United States tend to conceptualize race and racism, however, make it difficult for them to get started on the kind of analysis Baca Zinn and her colleagues recommend. First, race must be reconceptualized as a relationship rather than a "thing" or inherent property of people. (Recollect the argument from an earlier chapter that there are no inherently masculine or feminine properties, that gender is a socially defined relationship.) Then it is easier to understand how racism functions: a form of it such as white supremacy appeals to purported biological differences in order to provide a rationale for the appropriation of material and social resources by one group at the expense of others.

Thus, second, racism is fundamentally a matter of social structure, not of individual bad attitudes and false beliefs. As sociologist David Wellman has pointed out, the tendency to define racism as "race prejudice" settles for an account that lodges responsibility for racism only on the already economically disadvantaged poor whites—the Archie Bunkers—who, in contrast to middle-class people, have not learned to avoid making overtly racist statements or have not been rewarded for doing so. Racism is enacted in many different ways, of which overt prejudice is just one. It is fundamentally a political relationship, says

30. Baca Zinn et al., "The Costs of Exclusionary Practices," 296–97. The references are to Palmer, "White Women/Black Women"; Bonnie Thornton Dill, "On the Hem of Life: Race, Class, and the Prospects for Sisterhood," in *Class, Race, and Sex,* ed. Amy Swerdlow and Hanna Lessingler (Boston: G. K. Hall, 1983); and Simons, "Racism and Feminism."

Wellman, a strategy that "systematically provides economic, political, psychological, and social advantages for whites at the expense of Blacks and other people of color," and it is a dynamic relationship flexible enough to adapt to changing historical conditions.[31] I do not mean to suggest removing responsibility from individuals for their beliefs and behaviors; it is wrong to express racist prejudices. But to arrive at this understanding is only the beginning of grappling with what constitutes racist beliefs and behaviors—with the institutional race supremacy that "nonprejudiced" individual beliefs and behaviors support and maintain. Feminist readers will recollect that we never tolerated such a shallow and unhelpful kind of analysis of sexism. Eliminating sexist comments and misogynist attitudes is important, but it is even more important to transform the political, economic, and social institutions that support and maintain sexism and androcentrism.

These two points lead to the third: namely, that the social structures of race relationships are interlocked with gender and class systems. This linkage is partially responsible for the flexibility and adaptability of each system of exploitation and oppression: since their origins, each has been used to construct the others.

Appreciation of these three aspects of race and racism can help analysts of science and technology to avoid otherwise tempting tendencies. They can avoid the tendency to suggest that women and men of European descent are colorless and recognize instead that women and men of European descent also bear race. They can avoid the tendency to "study down" by focusing only on women of color when race is the issue. (We feminists of European descent have frequently done this, even though we have strongly decried the analogous dynamic of men "studying women" but avoiding the use of feminist insights to study themselves and their institutions: "Why don't they study themselves if they want to find out about sexism and androcentrism?" we complain.) Of course, it would certainly be an improvement if people of European descent knew more about people of Third World descent, their lives, histories, thoughts, and writings, and if we appreciated more extensively the costs of the racism and imperialism in their lives in which we are so often complicitous—intentionally or not. But "study-

31. David Wellman, "Prejudiced People Are Not the Only Racists in America," in *Portraits of White Racism* (New York: Cambridge University Press, 1977), 37.

ing down" is certainly not the only way for people of European descent to learn these things.

Many other shifts of vision will occur as conceptually and empirically more adequate conceptions of race and racism provide resources for analysis. For example, the attempt to integrate race issues highlights the necessity of an adequate class analysis. The story of the development of modern sciences and technologies is inseparable from a history of class relations, and the existing class analyses provide important resources for studies of race and science. To take another example, shouldn't one speak of majority "overadvantage" as the logical companion to minority "underadvantage" or "disadvantage"?[32] Once people of European descent, too, are seen as bearers of race, and race is perceived as a structurally maintained relationship, it will no longer make sense to leave invisible in our conceptualization what happened to that missing "advantage." A similar point can be made about thinking of ending racism as a matter of "social justice." The concept of social justice suggests that some groups have it and others don't; therefore, society should focus on how to get it for those who don't. It fails to provide stalwart defenses against the tendency of many people to blame the victim and think that the have-nots must somehow deserve their condition. It does not lead us to think about how some people probably get "too much justice"—that is, are unfairly favored by social institutions. It must be as unjust for some to be the overbeneficiaries of social goods as it is for others to be the underbeneficiaries, but few people notice this when thinking about how to rectify violations of "social justice."

I address these comments primarily to feminists—although left analyses that do not integrate gender and race issues, and race analyses that do not consider gender (though they do usually integrate class issues), are equally incapable of reaching their own goals. Since gender and race are used to construct class—a working-class person must be "assigned" specific race and gender activities, attitudes, duties, and responsibilities—one cannot understand how class functions at any historical moment without grappling with its permeation by the dynamics of gender relations, Western imperialism, and white supremacy. The integration of race, gender, and class analyses is necessary for each

32. See Peggy McIntosh, "Understanding Correspondences between White Privilege and Male Privilege through Women's Studies Work" (forthcoming).

of the groups that have focused on one of these topics to attain its own goals.

Important issues about the participation of Western sciences and technologies in racial agendas are close to invisible from the perspective of the "add women of color and stir" projects with which I began this chapter. Two of these are taken up later: Chapter 9 looks at the relationship between the development of sciences and technologies in the modern West and the de-development of the sciences and technologies in the high cultures of the Third World; Chapter 11 explores the relationship between experience and knowledge in producing such accounts.

This chapter has attempted to introduce a competency standard to measure the feminist analyses of science and technology produced by people of European descent. A feminist analysis cannot advance only the interests of the speaker and of women in her or his race, class, and culture; such an analysis would more objectively be referred to as a self-interested analysis. What makes it self-interested instead of feminist is not that women try to advance women's interests but that they succeed in advancing only some women's interests—only the interests of women of their own race and class—and therefore in advancing some women's interests at the expense of others'. Feminism is supposed to improve the conditions of women, but this kind of analysis could improve the conditions of the speaker's group only at other women's expense. It is not intentions that set the standard of success but the actual consequences of the analysis. Feminist analyses must bring to consciousness and open for *discussion* the origins, consequences, values, and interests that they carry. This is part of feminism's scientific project, not an optional addendum.

Of course, the conventional discussions of the sciences deserve an equally rigorous criterion of adequacy. What can be done to enhance the democratic tendencies within the sciences and to inhibit their elitist, authoritarian, and distinctively androcentric, bourgeois, Eurocentric agendas?

9

Common Histories, Common Destinies

Science in the First and Third Worlds

To read the standard histories of the birth and growth of modern science which are intended for other than historians' eyes—for example, texts used to train scientists and popular accounts—is to enter an "altered state of consciousness." In the magical universe of these accounts, "real science" and the modern world that it makes possible— or at least those parts of it in which Westerners take pride—spring up out of a few brilliant ideas and some hardworking genius's stubborn insistence on fiddling with pieces of wire and glass. Modern science, we are told, arose out of the dissatisfaction of a few astronomers and astrologers with the complexity of the Ptolemaic universe; out of alchemical projects to transform natural resources into more valuable ones; out of Galileo's interest in the technical needs of Venetian engineers; out of the rediscovery of ancient Greek and Egyptian mystical beliefs about the nature of the universe, and so forth. These influences appear to have no relationship to one another except as they meet in modern science. From them springs the modern world, that significant human achievement so much more advanced than either the medieval European world or the "primitive cultures" that were the only ones to exist in most of the rest of the world until well into the twentieth century. Reflective adults might well suspect that there is a more complex story to be told, but this one creates a kind of emblem, an image saturated with social meanings that are continually disseminated by both scientists and historians of science when they are not speaking only among themselves (and frequently even when they are).

These stories emphasize the important contributions that the North

Atlantic sciences and their technologies have made to "our" lives but rarely mention the down side of this "human progress." Nor do they address the long and historically varying relationship between capitalism and science, its very good consequences for elites in the West, and its less good and often very bad consequences for people in virtually every other group. When we look around us in the United States, we can see that if the modern sciences and their technologies have produced marvelously engineered modern cities with sensuously beautiful mirrored skyscrapers, in those very same cities something has brought about urban megadisasters: in monstrous slums people live in hovels with only minimal sanitation; African American male unemployment is over 40 percent in some age brackets; infant mortality is among the highest in the Western nations; and thousands of homeless men, women, and children live on the streets. "Underdeveloped" communities as populous and as immiserized as entire Third World nations live right in the middle of a half-dozen or so U.S. cities where the life worlds of the wealthy are supposed to represent the highest achievements of Western civilization—achievements that make us "civilized" in contrast to "primitive" and "underdeveloped" cultures.

Among the consequences of modern science and its technologies are men on the moon—but also stockpiled nuclear bombs. Supermarkets are bursting with picture-perfect fruits and vegetables, many of which have been grown on the other side of the globe—yet starvation, malnutrition, and technically easily curable diseases are devastating entire populations, entire races. Some people in the West walk around with others' hearts, lungs, and kidneys in their bodies—but passing them on the street, and elsewhere around the world, are millions of people without access to the most basic sanitation, nutrition, or health care. Dystopian science fiction novels can hardly show us more than we can see if we look around us.[1] To modern Western science is attributed responsibility for much of the good in the world, but if it is responsible for the good parts, why isn't it also responsible for the bad parts? The conventional accounts make incomprehensible the relationship between the full magnificence and full horror of late twentieth-century scientific and technological cultures.

1. The dystopian visions of the destiny of the modern West in, e.g., Ursula LeGuin, *The Dispossessed* (New York: Harper & Row, 1974), and Marge Piercy, *Woman on the Edge of Time* (New York: Fawcett Books, 1976), feel as close to documentary as they do to fiction.

Since Mary Shelley's Frankenstein story, novelists, philosophers, filmmakers, theologians, and other thinkers have enjoyed insisting that modern science is *fundamentally* either utopian or dystopian and have issued dire warnings from one side or the other of these supposedly exclusive alternatives. But to end the discussion there, with the mystery of the dichotomy and the warnings, stops short of showing that what appears to be a mysterious contradiction actually has perfectly evident causes if we choose to look at them.

I want to focus on important ways in which the standard story of science isolates the development of Western science's cognitive structures—the abstract laws of nature—from some of their material conditions and consequences. Why the world that science and its technologies has helped to make has these contradictions is not as mysterious as the conventional accounts and the utopian-dystopian debate suggest. Recent histories and sociologies of science have begun to dissipate the mystery, as we have seen in earlier chapters.[2] But they still leave its outlines relatively firmly contained within modern Western history. The conditions and consequences I want to identify are to be found in Africa, Asia, and all the Americas as well as in Europe. There is another story of the origins of modern science, its technologies, and the world it helps make possible.

This second story requires us to look at the relationship between the development of the North Atlantic cultures, in which science and its technologies have played such an important role, and the de-development of the societies of Africa, Asia, and other Third World countries. This other story can be pieced together from sources that cannot claim

2. E.g., Stanley Aronowitz, *Science as Power: Discourse and Ideology in Modern Society* (Minneapolis: University of Minnesota Press, 1988); Morris Berman, *The Reenchantment of the World* (Ithaca: Cornell University Press, 1981); Brian Easlea, *Witch Hunting, Magic, and the New Philosophy* (Brighton, Eng.: Harvester Press, 1980); Paul Forman, "Behind Quantum Electronics: National Security as the Basis for Physical Research in the U.S. 1940–1960," *Historical Studies in Physical and Biological Sciences* 18 (1987); Stephen Jay Gould, *The Mismeasure of Man* (New York: Norton, 1981); Margaret Jacob, *The Cultural Meaning of the Scientific Revolution* (New York: Knopf, 1988); William Leiss, *The Domination of Nature* (Boston: Beacon Press, 1972); Richard Levins and Richard Lewontin, *The Dialectical Biologist* (Cambridge, Mass.: Harvard University Press, 1987); Carolyn Merchant, *The Death of Nature: Women, Ecology, and the Scientific Revolution* (New York: Harper & Row, 1980); Hilary Rose and Steven Rose, eds., *Ideology of/in the Natural Sciences* (Cambridge, Mass.: Schenkman, 1979); Londa Schiebinger, *The Mind Has No Sex: Women in the Origins of Modern Science* (Cambridge, Mass.: Harvard University Press, 1989).

the scholarly legacy that helps to legitimate the Western accounts. These sources do not bring with them many footnotes to five centuries of testimony by Western scientists and observers of science as recorded in books, libraries, and doctoral dissertations—although they do bring a few. These sources definitely feel lighter in weight, more fragile, in their evidential underpinnings. It is as if they veer off the road at some moment or other and then move off on newly visible paths into other worlds where our narrative has never ventured. Our stories are left on their beaten track to which these side paths through the underbrush are only incompletely attached.

In favor of the second story, however, is an increasing amount of empirical evidence for many of its claims. Moreover, it gains plausibility as its narrative structure is more and more supported by similar revisions of other areas of Western and Third World history; it is supported by a general shift in Western conceptual schemes in the direction of the one framing this other narrative. These two sources of ballast permit the second story to confront the familiar one if not on equal terms, at least in an increasingly challenging way.

This chapter reviews the first and sketches out the second story, reflects on what this other story could or should mean to white Westerners (including feminist critics of science and technology), and raises cautions about some all too tempting ways for people of European descent to relate to this kind of narrative.

Our Story and an Other Story

We are all familiar with the story of the birth of modern science. In important respects its broad outlines also define the terrain for historians and philosophers of science. With few exceptions, the main events and processes in this story take place within the boundaries of modern European history as that history is understood in the North Atlantic societies.

In its older form, this story is an "internalist" history of science: it assumes that histories of intellectual structures can be independent of the histories of the economic, political, and social environments in which the intellectual structures emerge, take shape, change, and die out or are transformed. It assumes that the activities of minds—at least of certain kinds of minds—can achieve a significant degree of independence from the economic, political, and social activities of the bodies in which these minds have their historical locations. Thus, this kind of history seeks simultaneously to reconstruct the logical development of science and also provide a historical explanation for it. The logical development *is* what should count as the historical explanation, perhaps here and there embellished with a few anecdotes about the Great Scientists.[3] This is still the preferred form of the story for many philosophers of science and historians.

One might expect greater concern with the historical situatedness of the sciences and their technologies from the "externalist" historians of science and from sociologists, feminists, and other thinkers who argue that political, economic, and social environments constitute boundary conditions for the development of science both as a social institution and as a history and logic of thought.[4] But such expectations are by and large disappointed. Even those intellectuals who have provided the most widely respected external histories of science do not usually challenge the idea that the proper geographic and temporal boundaries for these stories are modern European history. They have looked at the institutions and thought of science from the perspective of class struggles,[5] of women,[6] and of the infrastructures of modern sciences,[7] but they have done so primarily only from within the conventional boundaries of modern North Atlantic history. One important exception to this generalization is Joseph Needham, whose studies of the influence of Chinese science on Western sciences began appearing thirty-five years ago.[8]

Three recent books by Third World scholars challenge the idea that Western science can be adequately understood within the constraints of conventional Western history (or, rather, two do, and the third can be used to do so). They make a persuasive case for considering Western

3. Wolfgang Van den Daele, "The Social Construction of Science," in *The Social Production of Scientific Knowledge,* ed. Everett Mendelsohn, Peter Weingart, and Richard Whitley (Dordrecht: Reidel, 1977), 27.

4. As Van den Daele (ibid.) puts the point.

5. E.g., Rose and Rose, *Ideology.*

6. E.g., Merchant, *Death of Nature.*

7. E.g., Bruno Latour and Steve Woolgar, *Laboratory Life: The Social Construction of Scientific Facts* (Beverly Hills, Calif.: Sage, 1979); Karin Knorr-Cetina, *The Manufacture of Knowledge* (Oxford: Pergamon Press, 1981).

8. See, e.g., Joseph Needham, *The Grand Titration: Science and Society in East and West* (Toronto: University of Toronto Press, 1969).

obliviousness to these constraints not as an innocent oversight on the part of Westerners but as an important piece of racist and imperialist ideology that has helped to justify to Westerners the inevitability of the scientific and technological gap between the First and Third Worlds.

Science and Technology in the African High Cultures

The essays in Ivan Van Sertima's collection *Blacks in Science* report recent discoveries of ancient African scientific and technological achievements that have been invisible to the West for several centuries.[9] I review the major findings in these papers in order to make clear that the issue is not the reevaluation to a higher, more scientific status of apparently magical, mystical, or intuitional ways of interacting with the world.[10] Instead, it is the development in Africa of precisely the kinds of scientific and technological innovation that have been regarded as marks of the superiority of North Atlantic cultures.

In the 1970s it was discovered that Africans living in Tanzania 1,500 to 2,000 years ago had produced carbon steel by a method of such technological sophistication that it was not matched in Europe until the mid-nineteenth century.[11] The ruins of an astronomical observatory dating back to 300 B.C., uncovered in present-day Kenya, indicate that those who built and used it possessed one of the most accurate of pre-Christian calendars.[12] In West Africa between 1200 and 1400, the Dogon reported the rings of Saturn, the moons of Jupiter, and the spiral structure of the Milky Way galaxy: "They knew a billion worlds spiralled in space . . . that the moon was a barren world. They said it was 'dry and dead, like dried blood.'" They also knew that a small star, invisible to the naked eye, had an eliptical orbit around the star Sirius that took fifty years to complete. The discovery of perfectly spherical and very precise crystal lenses that date back to the period when Egypt was occupied by "Africans" (before the eleventh-century

9. Ivan Van Sertima, ed., *Blacks in Science: Ancient and Modern* (New Brunswick, N.J.: Transaction, 1986).

10. This has been the illuminating project of such studies as Robin Horton, "African Traditional Thought and Western Science," pts. 1–2, *Africa* 37 (1967).

11. Debra Shore, "Steel Making in Ancient Africa," in Van Sertima, *Blacks in Science.*

12. B. M. Lynch and L. H. Robbins, "Namoratunga: The First Archaeoastronomical Evidence in Sub-Saharan Africa," in Van Sertima, *Blacks in Science.*

spread of Islamic culture) provides evidence for supposing that the ancient Africans had invented telescopes.[13]

Mathematics was at least as highly developed in ancient Africa as in any other contemporary culture. Records of an 8,000-year-old numeration system, perhaps used as a lunar calendar, have been found in Zaire.[14] Mathematics develops to resolve social needs—to count or measure in increasingly complex ways, for instance[15]—so it should not be surprising to find in African societies sophisticated mathematical systems that were developed centuries or even millennia ago to aid commerce, and ancient vast and complex engineering and architectural achievements of the African high cultures. The pyramids of Egypt are the most obvious example of such engineering feats on the African continent, but south of the Sahara can be found Great Zimbabwe, a massive stone city more than eight hundred years old that was not rediscovered by researchers until 1982; 10,000 people are estimated to have lived there. The remains of more than two hundred smaller but similar stone villages are scattered over Zimbabwe and Mozambique. And in southern Africa can be found the most ancient mines in the world.[16]

Extraordinary engineering skills can also be seen in the history of African transportation. Sophisticated boats of various design were used to transport traders along the marine highways of the Niger and the Atlantic coast. Commercial fleets of as many as eighty large boats have been reported, with an average vessel length of ninety to one hundred feet. Even on land, such skills as the use of compasses and of astronomical computations were necessary for travelers of the trade routes crossing the Sahara. Van Sertima explains:

> The journey across the Sahara . . . is twice as long and twice as hazardous as a journey [made] by Africans across the open sea from Africa to America (1,500 miles). Africans had to cross thousands of miles of

13. Hunter Havelin Adams III, "African Observers of the Universe: The Sirius Question," and Adams, "New Light on the Dogon and Sirius," both in Van Sertima, *Blacks in Science.*

14. Claudia Zaslavsky, "The Yoruba Number System"; and Beatrice Lumpkin, "Africa in the Mainstream of Mathematics History," both in Van Sertima, *Blacks in Science.*

15. See the discussion of this point with respect to Western cultures in David Bloor, *Knowledge and Social Imagery* (London: Routledge & Kegan Paul, 1977).

16. Beatrice Lumpkin, "The Pyramids: Ancient Showcase of African Science and Technology"; Molefi and Kariamu Asante, "Great Zimbabwe: An Ancient African City-State," both in Van Sertima, *Blacks in Science.*

trackless wasteland, whereas the Atlantic ocean has natural seaways (Heyderdahl calls them 'marine conveyor belts') which automatically direct the seafarer, whether on an accidental or planned expedition. Africans had to solve the problems of storing grain for months while traversing the barren sands, whereas the sea is a mobile foodstore. Africans had to take huge waterjugs in their caravans to ensure supply before the next oasis, while fishjuice supplemented rainfall, however uncertain, in the trade wind zones of the Atlantic.[17]

Agricultural sciences developed in Africa at least seven millennia before they appeared on any other continent. Barley and wheat were cultivated and harvested near the Nile and farther south in Nubia more than 10,000 years before the Egyptian dynasties. Cattle were domesticated in the Kenya highlands more than 15,000 years ago. It now appears that the Euphrates River Valley, familiar to North Atlantic schoolchildren as the "cradle of civilization," was developed through the diffusion from Africa of information, ideas, and technologies. Plausible arguments can even be made that 5,000 years before Columbus "discovered" America, Africans brought plants to this hemisphere.[18]

In the field of medicine, as Van Sertima demonstrates, African herbal remedies used many of the components of modern drugs, including those of aspirin, Kaopectate, antibacterial agents, and reserpine. Effective treatments were devised for "abortion, retarded labor, malaria fever, rheumatism, neurotoxic venoms, snakebite, intestinal parasites, skin ulcers, tumours, catarrh, convulsions, venereal disease, bronchitis, conjunctivitis, urethral stricture, and others." Africans conducted sophisticated autopsies, just one demonstration of their extraordinary surgical skills. Even in the 1870s, after centuries of Euroamerican destruction of African culture and technological skills, Westerners reported observing in East Africa a Caesarean section at a time when that procedure was rare in the West: "The skill demonstrated in this operation startled readers of the Edinburgh Medical Journal where it was reported. The Africans were not only found to be doing the Caesarean section with routine skill, but to be using antisep-

17. Van Sertima, *Blacks in Science,* 19; Stewart C. Malloy, "Traditional African Watercraft: A New Look," in Van Sertima, *Blacks in Science.*

18. Fred Wendort, Romuald Schild, and Angela E. Close, "An Ancient Harvest on the Nile"; and Bayard Webster, "African Cattle Bones Stir Scientific Debate," both in Van Sertima, *Blacks in Science.* See also Ivan Van Sertima, *They Came before Columbus: The African Presence in Ancient America* (New York: Random House, 1977).

tic surgery, which Lister pioneered only two years earlier than this
event, and when the universal application of his methods in the operat-
ing rooms of Europe was still years away."[19]

Finally, contrary to Western assumptions that traditional Africans
were preliterate, there is evidence of the invention of at least half a dozen
scripts in ancient Africa, some originating as early as 3,000 B.C.[20] This
is especially significant for the reevaluation of African science and
technology, since it is widely argued that critical thought needs writing
as a resource. Certainly, both ancient and modern Western science have
benefited from records of observations and reasoning that can be passed
down from generation to generation on paper, freeing the mind from
tasks of memory so that it can examine critically the evidence for the
best beliefs. The African high cultures, too, could make use of this
resource.

Van Sertima argues that the appearance of a legacy of African primitiveness is the consequence of the destruction wrought by the North
Atlantic slave trade. He asks readers to imagine what would happen if
the primary centers of contemporary North Atlantic science and technology were to disappear.

A nuclear war could shatter the primary centers of twentieth-century
technology in a matter of days. The survivors on the periphery, although they would remember the aeroplanes and the television sets, the
robots and the computers, the space machines now circling our solar
system, would not be able for centuries to reproduce that technology.
Apart from the almost wholesale slaughter of the technocratic class, the
interconnection between those shattered centers, and the equally critical interdependency between the centers and their peripheries, would
be gone forever. It would be like the strands of a web which once
stretched across the world, left torn and dangling in a void.

A dark age would certainly follow. Centuries afterwards, the technological brilliance of the twentieth century would seem dream-like
and unreal. Until archeology began to pick up the pieces, those of us

19. Van Sertima, *Blacks in Science*, 23. See also Frederick Newsome, "Black Contri-
butions to the Early History of Western Medicine"; and Charles S. Finch, "The African
Background of Medical Science," both in Van Sertima, *Blacks in Science*.
20. Clyde-Ahmad Winters, "The Ancient Manding Script"; Willard R. Johnson, "The
Ancient Akan Script: *Sankofa* by Niangoran-Bouah," both in Van Sertima, *Blacks in
Science*.

who follow in the centuries to come will obviously doubt what had been achieved in the centuries preceding the disaster.[21]

Europe's Underdevelopment of Africa

The creation of that dark age is the subject of the second book, Walter Rodney's *How Europe Underdeveloped Africa.*[22] Rodney argues that African underdevelopment today must be seen as the other half of the history of European development. The two must be examined together, as a totality, because each is explainable only in terms of the other. The development of Europe occurred as a result in large part of the arrest of development processes going on in Africa; the increased scientific and technological literacy of Europeans was created in large part through the scientific and technological deskilling of Africans. Moreover, this causal relationship was not unforeseen. It was intended, claimed, by at least some Europeans from the very beginning and continued to be stated as official policy right up through the end of the colonial period in the 1960s. We could put Rodney's point another way by saying that the history of Europe and the United States *is* African history also and that African history *is* also European and U.S. history.

Joseph Needham pointed out in the 1950s that the West has frequently had no clear idea of where its intellectual and technological resources originated. He criticized leading Western historians of science for reflecting the prejudices of their culture in their studied ignorance about the influence of the ancient cultures on European science.[23] Rodney develops a related theme: the dependency of the North Atlantic nations on *economic and political* resources originating in Africa. Rodney's title identifies the gestalt switch required if we are to begin to gain a realistic grasp of this period of world history. It is not "African underdevelopment" that needs to be examined but the European activity that gave the "underdevelopment" of Africa its shape. Correlatively, our focus should be not on "European development" but on African contributions to the development of Europe. Needham's

21. Van Sertima, *Blacks in Science*, 8.

22. Walter Rodney, *How Europe Underdeveloped Africa* (Washington, D.C.: Howard University Press, 1982). This book was first published in London and Tanzania in 1972 and in the United States in 1974. Rodney was assassinated on June 13, 1980.

23. Needham, *Grand Titration*, esp. chaps. 2–3.

and Rodney's theme is one that an increasing array of thinkers are finding important as the North Atlantic nations lose their ruling position in the contemporary world. How is the history of Western Europe and the United States also the history of the Third World? How is Third World history also European history?[24]

Although Rodney's book is about general economic and social development, he does make some comments about science and technology, and we can draw inferences from other passages. The birth of modern science and the development of new technologies were necessary for the takeoff of industrialization in Europe. As Sal Restivo has put this point,

> the origin and development of modern science is inextricably intertwined with the origin and development of modern society. . . . The scientific revolution was one of an interrelated set of *parallel* organizational responses within the major institutional spheres of Western Europe from the fifteenth century onwards (including Protestantism in the religious sphere and modern capitalism in the economic sphere) to an underlying set of ecological, demographic, and political economic conditions. . . . [This] "parallel responses" thesis sets modern science into the very core of the modern state and its technological foundations.[25]

By the late thirteenth century, feudalism had become widespread in Europe, while only a few areas of Africa were beginning the transition

24. Among works that insist on the importance of seeing causal relationships between histories that in the West have been regarded as separate and self-contained, one early important study is Eric Williams, *Capitalism and Slavery* (Chapel Hill: University of North Carolina Press, 1944). Two later studies that stimulated much subsequent work (and controversy) are Immanuel Wallerstein, *The Modern World System: Capitalist Agriculture and the Origins of the European World Economy in the Sixteenth Century* (New York: Academic Press, 1974); and André Gunder Frank, *World Accumulation, 1492–1789* (New York: Monthly Review Press, 1978). See also Eric Wolf, *Europe and the People without History* (Berkeley: University of California Press, 1982). Other critics have analyzed the development of Western myths, ideologies, and silences that justified or hid the actual relationships between the West and "the Rest." See, e.g., Samir Amin, *Eurocentrism* (New York: Monthly Review Press, 1989); Martin Bernal, *Black Athena: The Afroasiatic Roots of Classical Civilization* (New Brunswick, N.J.: Rutgers University Press, 1987); Toni Morrison, "Unspeakable Things Unspoken: The Afro-American Presence in American Literature," *Michigan Quarterly Review* 28:1 (1989), which includes a startling rereading of Melville's *Moby Dick;* V. Y. Mudimbe, *The Invention of Africa: Gnosis, Philosophy, and the Order of Knowledge* (Bloomington: Indiana University Press, 1988); Edward Said, *Orientalism* (New York: Random House, 1979).

25. Sal Restivo, "Modern Science as a Social Problem," *Social Problems* 35:3 (1988), 210–11.

from communalism to feudalism. Rodney argues that this pattern of uneven development was responsible for placing Europe in a position to be able to create forms of science and technology that would, in the centuries to come, produce needs for labor and raw materials that could not be satisfied within Europe. Feudal economies provided fertile ground for the development of science and technology: science in Europe was developed by landowners—feudal nobles and, later, bourgeois farmers—who had a professional interest in the land.[26] From these landowners, a new social group began to appear that saw its own advancement in the linking of increased material riches, the control of nature, and the establishment of representative democracy. But in order to realize the first two of these goals, far greater resources in the way of raw materials and human labor were required than Europe could supply. Africa was one major place where such resources were spotted.

Slavery was certainly not invented by modern Europeans, nor did the enslavement of Africans begin with the North Atlantic trade. The ancient Greek civilization, so inspiring to the North Atlantic theorists of modern democracy, was a slaveholding society; as few as 10 percent of Athenian residents were eligible to count as citizens. Moreover, African tribes had already enslaved each other, and an Islamic-African slave trade had existed for centuries. Nevertheless, the North Atlantic trade had a particularly bad effect on African scientific and technological development and particularly good consequences for the development of science and technology in Europe and the United States.

Among the bad consequences for Africans was, most obviously, a huge population loss among the young, the healthy, and the child-bearers during the years when Europe's population was increasing fourfold.[27] Such a loss decreased productivity and weakened the ability of African cultures to tame and harness nature. A climate of social violence—between slavers and Africans, and between the African groups that slavers set against each other—discouraged attention to scientific thought and technological invention. Moreover, the flooding of European goods into African markets resulted in the loss of the technological skills necessary for the production of textiles and goods

---

26. Rodney, *How Europe Underdeveloped Africa,* 41.

27. Rodney (ibid., 96) suggests 10,000,000 as a conservative estimate of this loss. His evaluation is supported by Philip D. Curtin, *The Atlantic Slave Trade: A Census* (Madison: University of Wisconsin Press, 1969), cited in Wolf, *Europe,* 195.

in the indigenous economies. Production for export and the decline of population contributed to famine. Borrowing from other cultures, which was so important to the development of European science and technology, was greatly reduced in Africa by the rechanneling of so many transportation routes to the coasts, where the slavers' ships awaited their cargo.

Perhaps most important, Rodney argues, slaving created not just underdevelopment but also a reduction of development opportunity through the loss of young innovators, the loss of economic demand for products, the loss of resources, the loss of the spirit of innovation, and, as indicated, the loss of opportunities to borrow new ideas from neighbors, Europeans, and other peoples of the world.[28] He asks readers to imagine history had the situation of the British and any one of the African peoples been reversed: if millions of Britons had been put to work as slaves outside their homelands over a period of four centuries, and if continental Europe had also been enslaved, the level of scientific and technological development in England would have been low indeed. Consider the possible alternative historical scenarios: Newton's mechanics would more likely have been discovered by a Tanzanian, and Einstein's relativity theory by a Senegalese; modernity would have belonged not to Europe but to Africa. Perhaps we would have had physics different in significant respects from Newton's and Einstein's but equally powerful in other ways. Perhaps other sciences would have developed more quickly than physics and astronomy, leading to different conceptions of human relationships to nature. Perhaps Africans would today be arguing over whether European underdevelopment and the sorry state of the whites living in their streets should be attributed to the biological inferiority of pale-skinned peoples or to their inadequate cultural response to an inhospitable social and natural environment. (Of course, this last scenario assumes that they would have developed a destructive class system and imperialist agendas parallel to those that do obtain in the West, and there is little reason except a commitment to Eurocentrism to make such an assumption.)[29]

Rodney tells a similar story about the benefits that accrued to Europe from the colonization of Africa from 1885 to the 1960s. During

28. Rodney, *How Europe Underdeveloped Africa*, 105.
29. See Amin, *Eurocentrism*.

this period the European administration of African colonies moved to the African continent, permitting even greater imposition of European political, economic, social, and psychological initiatives on the African peoples. This is the period in which North Atlantic science experienced vast expansion, professionalization, and increased incorporation into state projects.[30] It seems reasonable to suppose that the simultaneity of this development in Europe and de-development in Africa was not mere coincidence.

The Case of the Indian Subcontinent

In the third book, *Aborted Discovery*, Susantha Goonatilake builds on Needham's work and on world systems theory to report on the history of scientific and technological achievements in the Southeast Asian area, achievements that go unexamined in Western histories.[31] He points out that they are neglected because of the Western assumption that the history of science within the West contains the history of all science. Goonatilake paints a gloomy picture of the possibility of reversing de-development in order to create a future of equal scientific and technological advances by the West and the Third World. He argues that policies in the West today ensure that science and technology in cultures at the periphery will remain inferior to those at the center of world political-economic systems. Third World scientists remain inextricably tied into intellectual and political economies that in a centripetal fashion continuously suck into the North Atlantic from the Third World the cutting edge of scientific and technological advances, a disproportionately large share of their good consequences, and a correspondingly small share of their bad effects. Third World scientists are led to speak and write primarily to and for an audience of Western listeners and readers; the intellectual and technological world system makes it unreasonable for them to be primarily interested in Third World audiences. Recent literature on so-called Third World development provides a good look at how the bad consequences of

30. Rose and Rose, *Ideology.*

31. Susantha Goonatilake, *Aborted Discovery: Science and Creativity in the Third World* (London: Zed Books, 1984). The theory of Frank, *World Accumulation,* and Wallerstein, *Modern World System,* is called "world systems theory."

Western sciences and technologies are disproportionately distributed to the Third World.[32]

The three works cited here are not the only recent studies that enable us to get a critical look at Western sciences and technologies from Third World perspectives. Two other literatures in particular must be mentioned. African philosophers have struggled for several decades to define a distinctively African philosophy.[33] Central in these discussions has been the issue of what role in African thought should be played by the "mental apparatus" of Western imperialism,[34] which is grounded in conceptions of scientific rationality, objectivity, and progress developed precisely to *distinguish* "civilized" Europeans from "primitive" Africans and other "lower" peoples. The issue for many of these thinkers is how people of African descent can gain access to the skill and knowledge developed in the West without having to suffer the bad consequences of their Western uses. One could say that they are asking the science question in African diaspora studies.

Recent feminist studies of women in "development" can also be used to generate critical perspectives on Western science and technology.

32. The world systems theory used in one way or another by all three writers is not without its limitations. It is too economistic, too deterministic; it gives little weight to government actions or to the efficacy of any politics other than class struggles classically conceptualized; it does not do much to identify economic, political, and social contradictions that might provide opportunities for positive change. In consequence, it is a downright depressing theory offering little hope to groups that might well be effective at generating positive changes. See, e.g., Theda Skocpol, "Wallerstein's World Capitalist System: A Theoretical and Historical Critique," *American Journal of Sociology* 82 (1977); Robert Brenner, "The Origins of Capitalist Development: A Critique of Neo-Smithian Marxism," *New Left Review* 104 (1987). (I thank Cynthia Enloe and Evelynn Hammonds for discussions of this issue, and Enloe for the citations.) Nevertheless, the theory is important exactly because it calls attention to the necessity of identifying the causes of conditions in the First and Third Worlds which lie outside those worlds. It is a valuable antidote to prevailing narrow and self-serving accounts. For two world systems theorists who appear to be making progress in working against the theory's limitations, see Samir Amin, *Delinking* (London: Zed Books, forthcoming); and Maria Mies, *Patriarchy and Accumulation on a World Scale: Women in the International Division of Labour* (Atlantic Highlands, N.J.: Zed Books, 1986). See also Maria Mies, Veronika Vennholdt-Thomsen, and Claudia von Werlhof, *Women: The Last Colony* (Atlantic Highlands, N.J.: Zed Books, 1988).

33. Good accounts are Paulin Hountondji, *African Philosophy: Myth and Reality* (Bloomington: Indiana University Press, 1983); Mudimbe, *The Invention of Africa;* and J. E. Wiredu, *Philosophy and African Culture* (Cambridge: Cambridge University Press, 1980).

34. Mudimbe's phrase.

Many have argued that Third World women regularly lose status and power when their societies are "developed" under Western direction (even when their brothers and fathers play significant roles in directing that so-called development).[35] At least one group of critics, however, contends that Western policies—those of the World Bank in particular—systematically and intentionally reproduce Third World peasant populations, a decline in the quality of whose lives is perceived to be the affordable cost of increased profit and economic control in the First World. Women are the special target of such policies, though all Third World peasant populations are susceptible to them. That is, peasants are not an anachronistic remnant of earlier economies; instead, Western bankers are intentionally *increasing* the numbers of contemporary peasants in the world, turning their labor away from supplying themselves and their children with adequate food, shelter, and health care and toward the production of cash crops that produce profit for the West. Then, of course, First World women must find ways to consume more and more of the products Third World women are forced to produce for export. Ironically, in this sense it is Third World women who are not permitted to reproduce themselves and First World women who are forced to do so.[36] From the perspective of this account, the causally connected exploitation of both Third and First World women provides both the model and the foundation for maintaining Western imperialism.

What White Westerners Can Learn from the Other Account

It is time to evaluate what the foregoing story about Western sciences and technologies should mean to Westerners. Obviously, there is much that people of European descent have not known about African and other Third World histories and about their relationship to the conventional histories of the West. The details will no doubt provide a fruitful terrain for research and struggle in years to come. My focus, however, is on ways in which this Other story can lead to some new

35. For a good review of this literature, see Susan C. Bourque and Kay B. Warren, "Technology, Gender, and Development," *Daedalus* 116:4 (1987).

36. Mies, *Patriarchy and Accumulation;* Mies, Vennholdt-Thomsen, and von Werlhof, *Women.* See also Cynthia Enloe, *Bananas, Beaches, and Bases: Making Feminist Sense of International Politics* (Berkeley: University of California Press, 1990).

insights in the philosophy and social studies of the sciences and their technologies.

Common Histories, Common Destinies

To try to explain the rise of science and technology in the West without referring to the Third World histories with which that rise is causally linked (and vice versa) can produce only partial and distorted accounts. Intentionality is not the issue here. For the purposes of a causal account, it is not relevant that many Western scientists and technological innovators have not intended to de-develop the Third World in order to develop their explanations or disseminate their innovations (though some have had just these intentions). What matters to a successful—an objective—explanation of the rise of North Atlantic science is that it be able to identify the causes of events and processes, whether or not human intentions are among these causes.

From such a perspective, we should refer not to the development of Europe and the underdevelopment of the Third World but to the *over*-development of Europe and *de*-development of the Third World. This terminology enables us to keep in mind the world systems that create these tendencies and to assign causal efficacy in appropriate ways. Similarly, we should talk about overadvantaged and de-advantaged peoples, rather than suggesting by the use of "underadvantaged" alone that the de-advantaged should themselves bear primary responsibility for their conditions, or that the causes of their lack of advantage are uninteresting or irrelevant to changing their situation.

Western culture has had to develop distorted conceptual schemes in order to create the impression that the histories of Africa and the North Atlantic can be kept separate. As one example, African American philosopher Lacinay Keita points to the false and West-advantaging beliefs that ancient Egypt was not really part of Africa, that ancient Greece was really part of Europe, and that Islamic culture is entirely Arabic in origin. On the contrary, there were continual minglings and interchanges between the Mediterranean, African, and Middle Eastern cultures from antiquity through the modern era. Western scholars, intentionally or not, justify the West's inflated self-image and its de-valuation of the achievements of Third World cultures through their constant refusal to acknowledge these connections. The philosophical thought of Plato and Aristotle and of the European medieval and

Renaissance periods is not the unique legacy of contemporary people of European descent; contemporary Africans can rightfully claim these philosophies as their inheritance also.[37] The Third World and the West share both their histories and their destinies. To say this is not to assert *identical* histories and destinies for First and Third World peoples but, instead, to recognize that their pasts, presents, and futures are inextricably linked.

Third World Achievements

Moreover, many scientific and technological innovations that people of European descent tend to assume are uniquely their inheritance have been made independently in the Third World—sometimes far earlier than in the West. Westerners need to rethink their stereotypes of "primitive cultures," which are associated with the assumption that all sciences deserving the name have been developed in the West.

People of Third World descent can take credit for helping to make possible the sciences and technologies of the North Atlantic. It is common to find acknowledgments to mothers and wives whose activities have made possible the noteworthy achievements of famous men, yet not all of these acknowledged activities were *intended* to result in noteworthy achievements or fame and fortune for sons, fathers, and husbands. Many were necessary activities expected or demanded of women, or these women were acting to please themselves or perhaps simply to survive. In these cases, making possible someone else's achievements was an unintended consequence. Similarly, it seems reasonable to hold that people of African and other Third World descent deserve a good part of the credit for making possible with their labor and lives, regardless of whether they so intended, scientific and technological innovations in the North Atlantic societies. If those of us whose ancestors were illiterate peasants in Europe during the scientific revolution think we can claim this legacy, why should not others whose work also made possible the modern Western sciences? And perhaps the desire of Third World peoples today to design sciences and technologies that do not have the bad consequences of those in the West will create the legacy to which future generations of the West will turn in

37. Lacinay Keita, "African Philosophical Systems: A Rational Reconstruction," *Philosophical Forum* 9:2–3 (1977–78). See also Bernal, *Black Athena*.

order to catch up to forms of "international development" that have originated in the Third World.

"Origins Stories"

The Western story about the birth of modern science is, among other things, what anthropologists call an "origins story." It tells Westerners a history that we are to think of ourselves as sharing: where we came from, how we differ from other cultures, what characteristics of individuals and of our culture are most highly valued, what principles we are supposed to follow in life. In this respect it is little different from the Homeric sagas, the Bible, or stories about George Washington that are told to second graders. Every culture has origins stories, and "The Birth of Modern Science" is a popular one in contemporary Western culture.

The first thing Westerners should note about the Other story, I think, is that it has the same functions. A colleague reported that she first became aware of Rodney's book when a black student told her that at their university, it was being discussed in self-generated study groups of African Americans. As Marx's *Capital* did for study groups in the 1960s, discussion of the Rodney argument serves as a consciousness-raising experience about other histories that can give the struggles in people's lives new meaning. It serves to support an oppositional consciousness; it is a way of explaining the Eurocentric and white supremacist agendas of modern Western society that makes more sense than the accounts of those agendas offered by whites' explanations of minority lives and public policy. The Other story can be not only an interesting and controversial account within scholarly circles—though it should certainly become that—but also a way for Third World peoples to redefine against the dominant Western stories an alternative account of who their ancestors were, what values have been important to them and should still be, and exactly who is responsible for the present conditions of their lives and the lives of the dominant groups. In these respects it does not differ at all from the dominant Western story.

But the two stories have conflicting moral evaluations of Westerners and Third World peoples. Each story explicitly or implicitly accuses the other group of moral inferiority. In the Western story, Third World peoples appear as primitives, as children, as barbarians and savages, as outside of history and culture, with no redeeming cultural achievements. They are an obstacle to "human progress" and a threat to the West unless they are supervised by Westerners. In the Third World story, Westerners are greedy, barbarous, savage, arrogant, self-centered, hypocritical, willfully ignorant, and generally dangerous to themselves as well as to decent human life.

We should expect the heat to be intense when defenders of the two accounts confront each other. It is not just the facts that are at issue, though it is on the facts that the scholarly dispute will overtly focus. After hearing both stories, to accept either one and retell it as one's own is to make oneself complicit, intentionally or not, with the moral evaluations it promotes. It is not just facts that are at issue but also the whole moral and cultural identity of Westerners and Third World peoples that is inseparable from our senses of ourselves as humans. This leads to the next point.

Science as a Contested Terrain

We cannot turn to science to decide between the accounts. In pointing to an implicit moral dispute, I do not intend to devalue the importance of examining and locating evidence relevant to the various hypotheses proposed in the Other story, weighing the evidence carefully, and so forth. This should prove to be fruitful and valuable research, and it has only barely been begun. However, the project is even more difficult than it might first appear. If the dispute to be settled were one between two Westerners about an issue that apparently had no moral implications for their own lives, we could expect to see attempts to produce sound arguments intended to reach a disinterested and value-free position based on observation and reason alone (of course, as I have been pointing out, far too many disputes appear this way to elite Westerners). The procedure to be used to decide between these two stories is less obvious. Certainly, documents can be more carefully examined, data resifted, new evidence gathered, fresh light thrown on old judgments; and this work will be welcome. But it appears futile to imagine that one can locate an impartial, dispassionate, disinterested, value-free perspective from which to judge between the two stories—at least at this moment in history.

The changes required to bring the Western story in line with the Other one would be so fundamental that it is reasonable to regard the stories as significantly incommensurable. In particular, attempts by European Americans to dismiss the Other story on the grounds that it is not consistent with "the best opinions" in the West will not convince the unconvinced, because the Western standards for best opinions are exactly what the alternative accounts put at issue. Anyone who might think that he or she could claim an impartial, dispassionate, disinterested, value-free perspective on these issues is not only deluded but has also already sided with the West and its standards for finding the "best opinion." Insofar as we European Americans are loyal to what it means to be a person of European descent, we are, directly or indirectly, either a victim or a beneficiary of the outcome of the dispute.

Here are some of the questions about conventional Western descriptions and evaluations of modern science that arise from the perspective of the accounts reviewed above. Should North Atlantic scientific and technological innovation be thought of as progress when it appears that the majority of the world's people have experienced as exploitative and oppressive the social complex of which science is an integral part? Perhaps one should speak of science and its technologies as creating progress for the few and regress for the many. Why should the period when North Atlantic advance parties were intended to and did prepare the way for imperialism be presented as the "age of discovery" or the "age of exploration"? Why not the "age of imperialism," "the age of genocide," or "the age of appropriation"? The language of discovery and exploration does have one virtue: in a naive way it makes perfectly clear that the imperialist projects were linked to scientific activity and justified on the basis of their scientific rationality. Should one attribute the scientific and technological innovations in Europe to the supposedly unique spirit of inquiry of Europeans or to their excessively warlike spirit?[38]

It is not just particular historical sciences and technologies but also exactly what Western cultures intend to refer to and to mean by "science and technology" that are put at issue by the Other story. Far from being outside of political struggles, as most North Atlantic schools of science assume, what gets to count as science and technology can be seen here as matters of conflict. The character and meanings of "real

---

38. See Jacob, *Cultural Meaning;* and Restivo, " Modern Science."

science" are *not* beyond politics; they are resources over which contesting groups try to gain control.[39]

This is so in two ways. The history of science and its meanings are contested in this struggle: there are "facts" to be discovered, interpreted, debated, and fitted into coherent and plausible accounts. However, the epistemology of science is also at issue: for instance, it is clear that only through what is referred to by some science theorists as the "intrusion of politics into science" can one see the problems with the conventional story. From the perspective of an antiracist politics, one can see that this account reeks with political origins, concepts, research designs, procedures of research, and consequences. Given the different frameworks of the two stories, defenders of each can claim that the other is partial and distorted and that politics is responsible for these flaws. Empiricist epistemologies, the ones that fit best with the image of the ideal institutions of knowledge-seeking as value-free and disinterested, do not help here; the positivist elements of empiricism that express "the spontaneous consciousness of science" are part of what is at issue.[40] What is needed is for North Atlantic philosophies, histories, and social studies of the sciences to develop and disseminate standards competent to distinguish between those claims about North Atlantic sciences and technologies that benefit Western elites and "less false claims." At present, it is difficult to locate such standards.[41]

No Corner on Rationality or Superstition

I have already referred to an erroneous contrast that makes it difficult to get a realistic picture of modern science: the contrast between the purportedly exemplary rationality of European cultures and the allegedly primitive and childlike superstitious, religious, mystical, pre-rational, alogical modes of reasoning characteristic of Third World cultures and the premodern cultures of the North Atlantic. At the height of colonial anthropology, French anthropologist Lucien Lévy-Bruhl described as prelogical the mentality that Westerners claimed to

39. Gerald Turkel alerted me to the value of discussing this topic in these terms in his comment on the version of this chapter presented at the Delaware Seminars in Women's Studies, University of Delaware, Newark, December 1987.

40. Roy Bhaskar, *Reclaiming Reality* (New York: Verso, 1989), chap. 4.

41. Chapter 6 discusses such standards in another context.

have discovered among Third World peoples. Many decades later, British anthropologist Robin Horton argued that Westerners overlook the important ways in which African traditional thought is far more similar to than different from Western scientific thought. Horton's analysis generated a lively discussion among anthropologists and philosophers, occurring as it did in the midst of the post-Kuhnian anguish over the apparently failed rationality of science, disputes between "naturalist" and "intentionalist" philosophies of social science, the political uprisings of the 1960s in Europe and the United States and the beginnings of independence in the African states.[42]

Theorists of Third World descent have argued that Horton got only half the story right. For example, philosopher J. E. Wiredu maintains that the problem with such well-intentioned accounts as Horton's is that their authors—primarily anthropoloists or other Western visitors to non-Western cultures—lack familiarity with conventional physics, chemistry, and other natural sciences. This ignorance has consistently made these visitors of European descent incapable of adopting the same kind of critical stance toward Western science, Western scientists, and Western thought in general that they bring to the beliefs and behaviors of Third World peoples. Wiredu points out that many contemporary scientists engage in religious practices and make assertions originating in religious texts which are quite as irrational, "prelogical," and superstitious as those they criticize. Moreover, though Westerners may feel themselves to be models of rationality when engaged, for instance, in discussions about the thought of early modern philosophers such as David Hume and John Locke, in fact they do not usually examine these philosophers' thought with anything like the critical stance they adopt toward non-Western thought. It is much easier to be critical about the unfamiliar than about the familiar, and scientific focuses and patterns of thought are as traditional for the North Atlantic societies as African views and assumptions are for Africans. Some Western philosophers, as Charles Mills recounts, have recognized the histrionic and ritualistic character of large stretches of Western thought that have been characterized as the pinnacles of rationality:

42. See Lucien Lévy-Bruhl, *Les fonctions mentales dans les sociétés inférieures* (Paris: Presses Universitaires de France, 1910), trans. as *How Natives Think* (London: Allen & Unwin, 1926); Horton, "African Traditional Thought"; Hountondji, *African Philosophy;* Mudimbe, *The Invention of Africa.*

Hume pointed out long ago that, whatever skeptical iconoclasm with respect to everyday beliefs philosophers may indulge in privately (or with their colleagues), "immediately upon leaving their closets, [they] mingle with the rest of mankind in those exploded opinions." Nor is this necessarily just a matter of expedient conformity with the unenlightened herd, for he admits in his own case that after a few hours at backgammon, when he tries to "return to these speculations, they appear so cold, and strain'd and ridiculous, that I cannot find in my heart to enter into them any farther."[43]

The account of North Atlantic science and technology that is emerging from the perspective of Third World lives challenges both the nature and distribution of the concept of rationality that has been central to the self-image of Western scientific culture.

Rethinking the Relationship between Science and Technology

Does the Other story confuse science with technology? Many readers of the accounts of the African high cultures would insist that they involve only applied sciences and technologies, not "science proper," and that there are virtually no interesting relationships between the development of Western science and conditions in the Third World at that time or subsequently. In this view, the term "science" should be reserved for something that has arisen only in the modern North Atlantic cultures. Such a critic would argue that "science" means only a set of abstract sentences (preferably those that are purely formal, mathematical) such as Newton's Laws or Boyle's Laws, produced by certain methods for which the paradigm is the method of physics. "Pure science" is description of nature irrespective of the meanings such descriptions carry for anyone and irrespective of the uses to which such descriptions are put. Thus, "pure science" itself can be described and evaluated apart from the external origins of the scientific agendas that generated these descriptions of nature, the political implications of the technologies used to arrive at or the social meanings of these descriptions, and the uses to which scientific claims are put.

An obsession with this kind of analytic separation between "pure

---

43. Charles Mills, "Alternative Epistemologies," *Social Theory and Practice* 14:3 (1988), 238; the quotations are from David Hume, *A Treatise of Human Nature*, ed. L. A. Selby-Bigge (Oxford: Clarendon Press, 1888), 216, 269.

science" and the technologies of science ensures a faulty understanding of the scientific and technological enterprise, for there are a number of ways in which science and technology are parts of a social continuum, in which they participate in each other's activities. To review the points made in Chapter 4, science makes possible certain subsequent technologies and applications, and technologies make possible new kinds of scientific thought by making possible new kinds of experiments and observations. Neither of these relations is conventionally thought to threaten the purported purity of science. But what about the fact that socially legitimated scientific problematics are usually (always?) responses to social needs that get defined as technological ones, and that social meanings of technologies lead us to new experiences of the world? Here the purity of science is threatened behind the back of science. Here we are dealing with general cultural influences on science, with the way science is situated in its historical cultures in spite of scientists' attempts to levitate it to transcendental projects of interest only to pure reason (whatever that is).

In order to grasp the import of the Other story for assumptions about the "purity" of the abstract sentences of science, we must create a kind of history and philosophy of science that cannot be accommodated by talking about relations between externalist and internalist or political and intellectual histories of science. The two kinds of science history are not so discrete. Instead, we need to ask how the funding patterns and political directions of science, as well as the general preoccupations of the age, get "inside" the concepts and claims of the natural as well as the social sciences. How do external influences become internal thought patterns? We have yet to understand how the technological needs of imperialism left their marks within the thought structures of North Atlantic science.

Having said this, I must add that of course there are plenty of technological innovations that owe little to science. From the development of implements for gathering and hunting to the invention of contemporary backpacks in which to carry babies, it is clear that technological innovation can proceed independently of anything that it would make sense to call science. After all, even dogs and apes invent tools. Nevertheless, the problematics of purportedly "pure" science are linked to perceived technological needs and to real technological innovations and consequences in much more intimate ways than "pure science" enthusiasts can admit. Science and technology can be discrete not because science is autonomous from technology (for it never is) but because technological innovation is sometimes autonomous from science—but "sometimes" is not "always." It is an empirical matter to determine when technological innovations are indebted to science and when they are not.[44] That understanding leaves open interesting questions about whether technological innovation has its own epistemologies, ones that are not derivative from theories of scientific knowledge.

The "Natives'" View of Western Sciences

With respect to epistemological considerations, it is clear that there is an asymmetry between the authors of the two stories. Western history and philosophy of science have been written primarily by people who in two ways are "natives" of the enterprise they chronicle. Many are former scientists, or science enthusiasts, who share the ethos of the sciences. Others are Westerners who, intentionally or not, share the ethos of the West, which highly values the character and perceived consequences of scientific rationality. Identification with the culture of North Atlantic science creates a predisposition to accept at face value precisely what the Other story questions. Philosophers of social science who are natural science enthusiasts always argue that an account from the perspective of the natives is incapable of providing precisely the kind of causal accounts that have proved so valuable in the natural sciences. Social scientists who "go native" in the sense of restricting their accounts to what the natives intended or could see are limiting themselves to interpretations of phenomena rather than seeking the causal explanations that qualify an account as a scientific one. Natives are experts about their culture in some ways but far from expert about it in other ways. Experts about some aspect of nature are not for that reason experts about how their own expertise is produced or what its consequences are.

In contrast, the Other story has been written by "outsiders within," who must develop a "bifurcated consciousness." Even within Western science's own prescriptions for science, those who command the largest observational field—and especially a range of particularly critical (alternative) evidence—are supposed to be highly valued. Thus, the

44. See Needham, *Grand Titration*.

logic of these accounts should lead us to give a plausibility edge to the Other story over the Western one.[45]

Gender Issues

No doubt there are other general insights to be gained by reflecting on these two accounts of Western science, but some are of special interest to feminists. For one thing, Third World women can claim their share of the glorious histories of Third World sciences and technologies. Because women frequently have had higher status and power in Third World cultures than they do in the First World, their share of the responsibility for the sciences and technologies of their high cultures may be larger than anyone familiar only with First World women's histories would expect. Going back to the beginning, one would note that "Woman the Gatherer," who made such important contributions to the development of technology, was not from the North Atlantic area. What contributions have Third World women made to the development of the sciences and technologies of their cultures since then? For example, were there significant differences between the kinds of scientific and engineering projects favored during the reigns of African queens and African kings?

Such questions lead to the understanding that women participate in and experience science and technology as Third World or First World women, not just "as women." Feminist accounts of science and technology tend to generalize from the experiences of white, North Atlantic, middle-class women to all women in ways that cannot be supported by the evidence. As Western sciences and their technologies have "progressed," Third World women have lost more than white Western women and have lost it differently. Since all the world's women (and men) share common histories and common destinies, it becomes especially important for First World women to ask how Third World women fare in the technology transfers of so-called development today. What do Third World women want and need from North Atlan-

45. See Chapter 6 for discussion of the "strong objectivity" necessary to theories of scientific knowledge if they are to be capable of evaluating alternative accounts of Western science and technology such as this one.

tic science and technology?[46] How are the benefits and losses of North Atlantic scientific and technological enterprises distributed among women by race, class, and culture? These kinds of concerns should be focused not only on uses and abuses but also on the content of scientific claims in biology and the social sciences. Especially pertinent are claims about the history and sociology of science and technology.

When one examines the social meanings of science and technology, they reveal a curious coincidence between masculinism and Euro-centrism. Many features that North Atlantic feminists attribute to stereotypes and ideologies of masculinity appear also as the target of African and other Third World critiques of Eurocentrism.[47] First World feminists, as well as Third World scholars, need to provide causal accounts that acknowledge the overlap and antagonisms between these two dichotomies. Why is it that characteristics thought of in Africa as Eurocentric are thought of in the West as distinctively masculine? Why are there such similarities between the distinguishing characteristics of femininity and Africanity? The answer to this must be a complex one. It must discuss both the sexual and racial stereotypes of the Other which have been constructed by white, Western men to justify their right to rule. Sexual meanings have been intertwined with racial ones: the conceptual frameworks of sexual and racial stereotypes are mutually constructing.[48]

What kind of epistemology will be able to recognize the importance of learning from the standpoint of women of Third World descent? If feminists of European descent begin by asking questions about both First and Third World sciences and technologies from the perspective of the lives of women of Third World descent, their conceptual framework should shift significantly from that found both in the First World feminist accounts and in those Third World accounts that have tended to begin by asking questions from the lives of men of Third World descent. First World feminists can learn to ask different kinds of ques-

46. See, e.g., Bourque and Warren, "Technology, Gender, and Development"; and Mies, *Patriarchy and Accumulation*.

47. I discussed this matter in some detail in *The Science Question in Feminism* (Ithaca: Cornell University Press, 1986), chap. 7.

48. See, e.g., Sander L. Gilman, *Difference and Pathology: Stereotypes of Sexuality, Race, and Madness* (Ithaca: Cornell University Press, 1985); Nancy Stepan, "Race and Gender: The Role of Analogy in Science," *Isis* 77 (1986).

tions by starting their thought from accounts that begin in Third World lives. Third World critics are less loyal to the West than are First World feminists, who can learn additional ways to be "disloyal to civilization," in Adrienne Rich's words.[49] North Atlantic feminists can learn how to center issues of racism and imperialism while maintaining concern for gender. Can it also be a feminist stance sometimes to refuse to center gender, to federate with Third World women on issues of racism and imperialism even in those cases where gender may not be the primary issue?

This epistemological project raises questions about how feminists of European descent should go about starting thought in the lives of women of Third World descent. Feminists of European descent cannot think through either African or North Atlantic sciences and technologies "like an African"; we have not experienced the lives from which these accounts emerge. Moreover, our lives have frequently benefited from the very events and processes that have disadvantaged the lives of Third World women and men. And yet these other accounts can inform our thought in many ways. (Chapter 11 addresses this issue).

How should people of European descent center in their thought the kind of story this chapter has been considering about the relationship between the rise of modern science and technology in the North Atlantic communities and the de-development of Third World sciences and technologies? In light of the history of racist and imperialist First World thought about the Third World, it is easy to imagine inappropriate ways to do this.

For one thing, attention to African and other Third World histories could replicate the Eurocentric fascination with the exotic and do so at the expense of serious concern for the scientific and technological situation of Third World peoples today. Rooting about in the history of science and technology will not change what happened, a critic might say, and it distracts people from urgent political issues such as considering what is to be done now.

Or the focus on the scientific and technological achievements of Third World peoples before and outside of First World science could be

49. Adrienne Rich, "Disloyal to Civilization: Feminism, Racism, Gynephobia," in Rich, *On Lies, Secrets, and Silence: Selected Prose, 1966–78* (New York: Norton, 1979).

used to advance a more subtle form of just that preoccupation with the primitive that Van Sertima criticized. After all, voyages across the Sahara, plant breeding, and Caesarean births are still fairly primitive achievements if we compare them with trips to the moon, contemporary genetic interventions, and heart transplants.

Some readers may think that I am recommending a search for individual scientific and technological foremothers in Africa and other Third World societies. As feminists in the West have done in their own histories, am I seeking Great Women in the history of African and Asian sciences and technologies, figures who have made distinctive but overlooked contributions to scientific and technological innovation? Being preoccupied with them would establish the individuality of a few Third World women at the expense of appreciating the contributions of the masses whose efforts made those achievements possible. It would further advance North Atlantic preoccupations with individualism, with the "genius story" of the history of science and technology, at the cost of understanding the more collective and community-organized ways in which science and technology have been practiced in cultures that do not share this ethic of individualism.

Some could perceive this call for a specifically feminist approach to First World science's embeddedness in world history as seeking to divide and conquer Third World critical perspectives on North Atlantic society by claiming the activities and struggles of Third World women for feminism while ignoring, silencing in our discourse, those of Third World men.

Finally, and most generally, one could ask whether such a feminist approach would launch a second attempted colonization of Third World science and technology by one part—the feminist part—of North Atlantic culture. How can we white Western feminists read African history, for example, without projecting into it our own fantasies and desires? How can this effort avoid expanding the empire of feminist analysis while forgetting the pressing needs of Third World women today?[50]

Such perils threaten the project I am proposing at every step, but I do

---

50. See, e.g., Gayatri Chakravorty Spivak, *In Other Worlds: Essays in Cultural Politics* (New York: Methuen, 1987); and Spivak, "Three Women's Texts and a Critique of Imperialism," in *"Race," Writing, and Difference,* ed. Henry Louis Gates, Jr. (Chicago: University of Chicago Press, 1986).

not think them sufficient reason to pursue feminist—or "pre-feminist"—science and technology studies within a history of the field that appears so patently Eurocentric when considered in light of these accounts from the perspective of Third World lives. Eurocentric assumptions are "subjective" folk traditions of just the sort against which science is supposed to struggle. We can provide less partial and distorted understandings of First World science and its technologies if we can bring the rising contemporary skepticism about Eurocentric perspectives to bear on this social tradition, too. The Third World's other story opens up useful ways to problematize our identities as North Atlantic feminists—to be "disloyal to civilization"—in one more illuminating and politically useful way. Success in that attempted disloyalty, however, will require innovative strategies to avoid the perils that await the courageous.

10

Thinking from the Perspective of Lesbian Lives

Feminist standpoint epistemologies direct us to start our research and scholarship from the perspectives of women's lives. Earlier chapters have noted that class and race make important differences in women's lives; class, race, and gender are used to construct one another. But what about sexuality? Shouldn't there be a distinctive lesbian epistemological standpoint?[1] If so, what can it contribute to the natural and social sciences?

Rather than trying to survey all the literature or issues here, I want to examine the kinds of resources for research and theory that can be gained by taking seriously the directive of feminist standpoint epistemology to start thought from *all* women's lives, not simply from the lives of men in the dominant groups or even primarily from the lives of those women who are most highly valued by conventional androcentric,

---

1. See Chapters 5–7 for the "logic" of standpoint theory. Should I be referring to *a* lesbian standpoint, or to lesbian standpoint*s*? The former risks the appearance of essentialism; the latter risks confusing "the perspective from lesbian lives" with lesbians' experiences and statements. Individual experiences and testimony about them constitute the necessary starting point for developing explanations about anything in the world at all. But the theories developed as a result lead to reinterpretations of the experiences, and this is true for the individuals who have the experiences as well as for those who only hear about them (e.g., the theory that gays and lesbians are neither morally sinful nor biologically diseased but instead are politically oppressed has radically changed the way gay and lesbian people understand their own past and future experiences). I have preserved the singular in order to block the tendency to overvalue individuals' experiences or statements at the expense of theoretically mediated "objective accounts" (see Chapter 6) of the world from the perspective of their lives.

white, Western, economically advantaged, heterosexist thought. I intend in this chapter to continue countering the idea that there is some essential or typical or preferred "woman," from whose typical life feminist standpoint theory requires us to start. On the contrary, it is all women's lives—the marginalized as well as those closer to the center— from which we must look at social relations.

Many readers may think that there is something odd about having a chapter on "lesbian lives" in a book on the sciences and theories of knowledge. But two decades ago it would have been peculiar to have a chapter on women's lives in a similarly focused book. As noted in earlier chapters, feminism joined other social liberation movements in undermining the legitimacy of thinking in terms of universal "man." Subsequently, gay and lesbian movements, poor people's movements, movements of people of Third World descent, and others challenged the idea that any feminist thought should center on universal "woman." And since it is unlikely that readers who are unfamiliar with recent feminist social theory will have the slightest idea what it could mean to "start research from the perspective of lesbian lives," it is worthwhile to review some contributions to feminist theory and research that have already been made through just such a procedure.

Several points should be clarified at the outset. First of all, what is a lesbian? Most people probably feel pretty sure they know—until they pause to reflect on issues that have been raised in some recent writings. Must a woman have sex with another woman in order to be counted as a lesbian? Many people would think so, yet Adrienne Rich has argued that one should think instead of a "lesbian continuum" that includes all women who have engaged in resistance against compulsory heterosexuality, whether or not they have actually had sexual relations with another woman. Rich is emphasizing the importance of the political content of lesbianism.[2] Further, historians debate what should count as "having sex." Should shared beds, intimate touching, holding hands, kisses on the lips, and expressions of undying love, devotion, and passion—the characteristics of romantic friendships between women who were regarded by one and all as models of heterosexual womanhood in the nineteenth and early twentieth centuries—be sufficient

2. Adrienne Rich, "Compulsory Heterosexuality and Lesbian Existence," *Signs* 5:4 (1980).

evidence to conclude that these women had sexual relationships?[3] Is it fair to think of any of these women as lesbians when many of them were apparently also as happily married as most other married women of their day, when they did not so name themselves, when the more complex (and in many respects misogynist) analyses of sexuality by Freud and the sexologists had not yet become widely known, and when there was not yet the urban culture of economically independent women that has made possible today's self-consciously lesbian cultures?[4]

The past is not the only contested zone in thinking about sexuality. Is intentionally and self-consciously having sex with women sufficient to earn a woman of today the label "lesbian"? Some argue that it is not, since some women who have sexual relations with women also have ongoing sexual relations with men; bisexuality is different from lesbianism. Moreover, sometimes women engage in sex relations with other women, or appear to do so, precisely for the benefit of patriarchal culture, as in the so-called "lesbian photo" centerfolds of such magazines as *Hustler* and *Playboy*. In these, patriarchal culture's fantasies about female sexuality are staged as a spectacle to satisfy male desire. Why are they "lesbian" photos? To take another case, is it fair to categorize as lesbians women who themselves reject the name? Some women who appear to "live as lesbians" reject the name because they find lesbian feminism too radical or threatening to their otherwise conventional life-styles. Others reject the name because they think it too conservative a label for their lives and politics; these activists in the gay liberation movement or other "left" politics wish to disassociate themselves from the separatist or class-privileged or ethnocentric agendas in some lesbian tendencies that they find counterproductive.

Rather than entering these debates, I simply take one of several possible reasonable positions in this discussion: I shall count as lesbian all those women who have adopted the term for themselves. This includes too few women by some standards and too many by others, and it has consequences for what is meant by "starting thought in lesbian lives." But it does have the virtue of privileging an autonomy

3. Lillian Faderman, *Surpassing the Love of Men: Romantic Friendship and Love between Women from the Renaissance to the Present* (New York: Morrow, 1981).

4. Ann Ferguson, "Patriarchy, Sexual Identity, and the Sexual Revolution," *Signs* 7:1 (1981); see also Ferguson, *Blood at the Root: Motherhood, Sexuality, and Male Dominance* (Winchester, Mass.: Unwin Hyman, 1989).

for lesbians to name themselves and their worlds as they wish, an autonomy that women—and especially marginalized women—are all too often denied. The right to define the categories through which one is to see the world and to be seen by it is a fundamental political right.

In identifying what one can see with the help of a lesbian standpoint, I do not point exclusively to insights *about* lesbians. The standpoint epistemologies have a different logic. Just as the research and scholarship that begin from the standpoint of women more generally is not exclusively *about* women, so these insights are not exclusively about lesbians. The point is that starting thought from the (many different) daily activities of lesbians enables us to see things that might otherwise have been invisible to us, not just about those lives but about heterosexual women's lives and men's lives, straight as well as gay.

Nor have these insights necessarily been generated only *by* lesbians. Some men have clearly been able to think—at least occasionally— from the perspective of women's lives rather than from the immediately available understandings of their own lives. John Stuart Mill, Karl Marx, Frederick Douglass, and other male feminist thinkers have been able to generate original understandings from the perspective of lives that were not their own, or at least to use those perspectives to think from their own lives in radically new ways. Similarly, anyone who knows enough about them should be able to think from the perspective of lesbian lives. After all, we are all expected to start from the lives of people in radically different cultures when we are asked to explain the thought of Plato, Aristotle, Descartes, or Shakespeare. Contemporary women have been expected to be able to appreciate the world view of notoriously misogynist men—for example, Henry Miller and Norman Mailer (or Aristotle and Descartes, some critics would say). It should not be much more difficult for heterosexuals to think from the perspective of lesbian lives. Not all the people whose work I cite below have identified themselves as lesbian, and some identify themselves as heterosexual.

Finally, it needs to be mentioned that at least a latent love of women deeply permeates most (all?) feminist thought—just as the misogynists greatly fear—though it is not clear that all women who claim the name lesbian have what others would identify as a love of women. However, the fact of the feminist love of women can be confusing to onlookers who are not used to the idea of loving and valuing women for themselves rather than primarily for how they serve the needs of men, children, or the dominant groups in society. To such people, a love of women appears to be a betrayal of "the natural order"—that is, of patriarchal principles, or class loyalties, or racial pride, or cultural identity—and who violates such principles, loyalties, pride, or identity more blatantly than lesbians? or so their reasoning apparently goes. Feminism begins with a sense of moral outrage at how women are treated in both word and deed. In rejecting the sexist and androcentric thought that degrades and devalues women, it necessarily rejects the misogyny, the woman-hating (what Freud referred to as the "normal misogyny" that men feel toward women) that lies not so deeply veiled behind it. Thus, as women learn to "love themselves" in ways that a sexist and androcentric society forbids, they are led also to reevaluate the misogynist attitude that women, too, are expected to have toward women generally—toward other women as well as themselves. How could one learn to love oneself and people who are like oneself without also becoming aware of the eroticism of "our kind of people"? (I return below to some insights about male "homosociality" and female sexuality that lesbian perspectives make possible.)

These issues are far too complex to deal with here. I intend these preliminary remarks simply to unsettle conventional confidence about the usefulness of stereotypical views of women, sexuality, feminism, and lesbianism.

Contributions to Feminist Thought

Social analyses generated from the perspective of lesbian lives have greatly enriched lesbians' understandings of lesbian lives, of course, but my concern is a different one: what are the contributions to feminist thought more generally that taking the standpoint of lesbians enhances or generates? Without attempting a comprehensive literature review, I can nevertheless point to some striking insights that emerge from such a project.

(1) From a lesbian standpoint one sees women in relation to other women—or at least not only in relation to men and family. Literary critic Bonnie Zimmerman argues that "lesbians brought female bonding to the center of feminist discourse, and now most feminists see women in relation to other women." In contrast, she says, men and heterosexual women tend to focus on women in other ways.

Generally, men see women in relation to themselves as sexual objects or domestic servants. Throughout history, men have also seen women as exemplars or archetypes, both positive and negative: Eve or Mary, witch or saint, angel in the house or unsexed woman. Thus men see women either as appendages or as a class, but not as individual and independent persons with agencies and perspectives of their own.

Some [heterosexual women] state proudly that they do not see women at all ("I prefer talking with men"; "sex has nothing to do with my work or life"). . . . Another way heterosexual women see other women, according to patriarchal mythology, is as rivals. . . . Finally, a heterosexual woman may see other women within the roles and institutions established by a male-centered perspective: i.e., woman as wife, as mother, as seductress, as mistress, even as independent woman.[5]

Feminists have learned to see women in their relationships with one another as mothers (not just as mothers to their sons), daughters, sisters, lovers, friends, comrades in struggle, teachers, students, mentors, muses, co-workers, and colleagues. Historian Bettina Aptheker reports that it is exactly women's mutual love and support that has frequently provided the bedrock of their social activism *for* women. For example, the intense commitments between many of the women who were responsible for founding settlement houses and women's colleges made possible the heroic struggles to establish these institutions.[6]

Of course, not all relations between women have been inspiring ones. The new scholarship has shown the importance of looking at the relations between women that are enacted in racist or classist or imperialist conditions—between women and their slaves or domestic servants, between Aryan and Jewish women in Nazi Germany, between women of European descent and those "others" they helped to rule in Africa, Asia, and Latin America. Nevertheless, to choose to work specifically for women, to improve the conditions in which women live their lives, is to take a fairly sure risk of losing the approval of many men—men of the left as well as of the right. Women's work for women has had to be grounded in women's love and support for one another. Where else could it find a foundation?

5. Bonnie Zimmerman, "Seeing, Reading, Knowing: The Lesbian Appropriation of Literature," in *(En)Gendering Knowledge: Feminists in Academe,* ed. Joan Hartman and Ellen Messer-Davidow (Knoxville: University of Tennessee Press, 1991).
6. Bettina Aptheker, *Tapestries of Life: Women's Work, Women's Consciousness, and the Meaning of Daily Life* (Amherst: University of Massachusetts Press, 1989).

A related subject that is brought to the fore by thinking from the perspective of the daily lives of lesbians is single women's roles in history. Many lesbians have lived in conventional families, but many have not. The sexist and androcentric perspective that insists on seeing women primarily in conventional family settings cannot detect the social importance or the meanings to women of the work that single women have done. (Are all women who do not live in whatever counts as "normal families" in fact perceived to be "single" and "living alone" by conventional standards— "single parents," women who live in communes, dormitories, with female friends, and so on?) Most women in the North Atlantic countries are single for a good part of their lives. And where there have been large demographic imbalances between men and women (after major wars, or "back home" when the men have headed for the frontiers or the colonies), the number of single women has reached proportions never countenanced by those who insist that women's lives must be understood primarily within families. Looking at the world from the perspective of many lesbians' lives today brings into sharp relief the pains, pleasures, and achievements of single women's lives.

Could these contributions to feminist thought have arisen by starting from heterosexual women's lives? Theoretically, yes: clearly, heterosexual women today have important relationships with other women, and single heterosexual women make important contributions to the worlds around us. Nevertheless, in the conceptual frameworks structuring the history and sociology of heterosexual worlds, these aspects of women's lives have seemed less important, or more difficult to see, than women's relationships to men and as married women. Perhaps it has even seemed disloyal to the men upon whom heterosexual women have had to depend for economic and social support for women to dwell too long on the great value and obvious pleasures of women's lives spent with other women and outside of marriage. (Of course, men never think it disloyal to women to expect us all to dwell on the value and pleasure of *their* relationships with other men and outside of marriage; those relationships are simply called "science" "society," or "the social order.") Because lesbian lives tend not to center relationships with men let alone activities within heterosexual marriages, female bonding and single women are brought into sharper relief. But the value of a lesbian standpoint does not have to depend on this claim alone, well supported as many think it to be.

(2) A lesbian standpoint permits us to see and to imagine communities that do not need or want men socially. The perspective from men's lives enables us to see plenty of communities that have not needed or wanted to include women. Men have excluded women from many areas of their activity: laboratories and science directorates, the military, realms of adventure and exploration, intellectual circles and literary lineages, unions, bars, street corners, sports, the priesthood, higher education, university faculty, and more. Male homosocial worlds are the norm for men, especially in the upper and lower classes today. It is not really accurate, however, to say that these worlds have not needed women, for someone had to be doing the daily maintenance of men's bodies in order for men to carry on their other activities; this work was most often done by women.

It is more difficult to see communities of women that do not need or want men. "Women only" communities have often been directed or supervised by men: consider convents, women's colleges, single-sex paid labor (for example, in textile mills and secretarial pools), "ladies auxiliaries" of men's organizations, and so on.

From the perspective of lesbian lives, however, communities of women designed, organized, and directed by women become imaginable.[7] In heterosexual worlds, women's energies *for* women are distracted, undermined, and devalued by demands that they serve first, and frequently only, fathers, brothers, husbands, sons, and male lovers, colleagues, bosses, and comrades. To work *for* women—and not just in raising up future wives, in footbinding, performing cliterodectomies, and the like, but really *for* women—is to risk severe male disapproval. Charlotte Bunch's phrase "woman-identified-woman" carries very different meanings from its apparent structural analogue, "man-identified-man."[8] A woman-identified-woman suggests a woman who is a traitor to male supremacy and is probably going to be severely punished by the social order for her impudence in presuming her right not to be "for men." A man-identified-man suggests the norm for masculinity; what could a woman-identified-man be but a "queer"?[9]

---

7. See Zimmerman, "Seeing, Reading, Knowing."

8. See Charlotte Bunch, *Passionate Politics: Essays 1968–1988* (New York: St. Martin's Press, 1988).

9. Judith Roof points out that the displacement of the gay man into femininity is a way of assuring the "innocent" nature of the male homosocial bond among presumably heterosexual men (personal conversation.)

Another aspect of this issue has been brought out by Bettina Aptheker. She argues that a lesbian presence provides a clear sense of the
potential for a kind of female independence that is invisible from the
perspective of heterosexual women's lives. From that perspective, wom-
en can be seen primarily as valuing, circling around, and servicing
men. They must make the choice that psychologists have identified
between autonomy and attachment. For heterosexual women the
choice usually goes to attachment, but whichever way it goes, a choice
must be made. The presence of lesbians, however, changes the grounds
on which heterosexual women juggle the choice between autonomy
and attachment.[10] Aptheker is arguing that thinking from the perspective of lesbian lives makes visible the fact that women need not make
this choice, that it is possible for autonomy and attachment both to be
cultivated and to be cultivated as supports—as grounds—for each
other. Lesbian lives set new standards for what should be possible for
women within heterosexual relations.

Many women are beginning to find it not quite so necessary to their
fulfillment to have to live with a man. The predictability of women's
getting a fair exchange in trading sex and social services for economic
support—historically a compelling reason for many women to mar-
ry—is more undependable than ever. Of course, in Black and Hispanic
communities in the United States, racism and class exploitation ensure
that living with a man will by no means always raise a woman's access
to economic resources.[11] There must be institutional transformation
before poor men as well as women will be able to support themselves
and contribute to the support of kin. Men who do have access to
economic resources must become less dependent on women's labor in
their daily lives and more nurturing of women and children if they are
to be welcomed as more than an economic benefit of dubious reliability in women's lives.

(3) A lesbian standpoint reveals that woman (heterosexual) is made,
not born. Simone de Beauvoir declared that "woman" is a social construct, not a biological fact. But so too is compulsory heterosexuality,
as the perspective from lesbian lives has made clear. Appropriate heterosexual behavior varies from culture to culture and in different his-

10. Aptheker, *Tapestries of Life*, 93.
11. Maxine Baca Zinn and D. Stanley Eitzen, *Diversity in American Families* (New
York: Harper & Row, 1987).

torical eras, as anthropologists and historians have shown.[12] But beyond cultural variation in what counts as heterosexuality and
homosexuality lies the construction of compulsory heterosexuality itself.[13] As Gayle Rubin pointed out in her widely cited study, Freud and
Claude Lévi-Strauss had already in different ways clearly perceived and
described compulsory heterosexuality as a social construction, but the
political implications of this fact for women's lives and for feminism
were not detected until the second women's movement. Rereading
Lévi-Strauss from the perspective of lesbian lives, one can see that
principles of compulsory heterosexuality restrict women in ways that
they do not restrict men. Kinship systems are constructed through gift
relationships between men. Women, along with cows, shells, and
yams, are property that men give each other to make kin relations. A
gift cannot give itself; it requires a giver. Consequently, women's sexual
agency must be restricted by men if women are to function effectively
as the most prized objects in which men traffic.[14] It is only from the
perspective of the everyday lives of lesbians that one can see heterosexual privilege at all. From the perspective of heterosexual women's lives,
this privilege appears simply as "the way things are," perhaps as part of
nature.[15]

Rereading Freud from the perspective of lesbian lives, one sees that
everyone's first and most complete love is a woman—the mother. Men
get back as adults the kind of lover they had to give up as an infant—a
woman. But women get "only a man," who calls forth far less of the
pleasurable and painful memories of the "first love" than wives do for
husbands. Men's experiences of sexual relations with women consequently tend to be far more emotionally intense and textured than are
their partners'. As Catharine MacKinnon has put the point, only half-

12. See John D'Emilio and Estelle Freedman, *Intimate Matters: A History of Sexuality
in America* (New York: Harper & Row, 1988); Faderman, *Surpassing the Love of Men;*
Carroll Smith-Rosenberg, "The Female World of Love and Ritual: Relations between
Women in Nineteenth Century America," *Signs* 1:1 (1975).

13. See, e.g., Marilyn Frye, *The Politics of Reality: Essays in Feminist Theory* (Trumansburg, N.Y.: Crossing Press, 1983); and Monique Wittig, "One Is Not Born a
Woman," *Feminist Issues* 1:2 (1981).

14. Gayle Rubin, "'The Traffic in Women': Notes on the 'Political Economy' of Sex,"
in *Toward an Anthropology of Women,* ed. Rayna Rapp Reiter (New York: Monthly
Review Press, 1975).

15. See Judith Butler, *Gender Trouble: Feminism and the Subversion of Identity* (New
York: Routledge, 1990), for a subtle and startling analysis of sex identity as part of
gender and gender as fundamentally a performative act, not an interior state.

facetiously, the question should be not why some women become lesbians but why all women do not.[16]

Other aspects of lesbian lives lead to a different set of issues for feminism about compulsory heterosexuality. Not all homosexual women have found unproblematic their identity as women and with women. When Freud pointed out that in a certain sense femininity is a horrible burden for all women, he was equating "femininity" with the passivity as "human agents" that women are supposed to acquire (I return to this passivity shortly). But some lesbians have never acquired that femininity; those for whom expectations of feminine behavior were particularly onerous actively resisted acquiring the passivity to which Freud referred. If they were not passive, were they women? and if not women, then what? Starting thought from lesbian lives raises new questions about relationships between sex identity, gender identity, and politics.[17]

(4) A lesbian standpoint centers female sexuality, and female sexuality as constructed by women. Women's sexuality becomes central in a variety of ways through the perspectives available from lesbian lives. First of all, it may be reasonable to say that only through the perspective available from lesbian lives can one see that women have any sexuality at all—at least in contrast to the perspective available from the lives of men or of heterosexual women of the dominant races and classes. From the latter perspectives, female sexuality is often seen as only a biological object: it is how species and various classes, races, or cultures reproduce themselves. It was how Aryans were to increase in numbers and, through sterilization and enforced abortions, Jews were to decrease; it was how black slaves were to be "manufactured" on U.S. plantations. Female sexuality may also be seen as an economic object. "Good women" trade it, and the children they bear, to men in return for economic support; "bad women" sell it for men's use in piecework fashion. Or, as indicated, female sexuality is a factory, part of the system of economic production.[18] It can further be regarded as a

16. Catharine MacKinnon, "Feminism, Marxism, Method, and the State," pts. 1–2, *Signs* 7:3 and 8:4 (1982).

17. Frances Hanckel alerted me to the importance of these issues. See Butler, *Gender Trouble*, for suggestions about kinds of politics that become possible once we start asking such questions.

18. Emily Martin, *The Woman in the Body* (Boston: Beacon Press, 1987), discusses the factory metaphor that gynecology favors for conceptualizing the female reproductive system.

political object, as when aristocratic families exchange their daughters to cement political loyalties.[19] It has been seen as a means of giving pleasure to men: popular magazines instruct their female audiences never to frustrate their lovers' desires; a "real woman" pleases her man—evidently regardless of her own fears or desires. From the perspective of heterosexual women's lives, women appear to have no sexual agency. It is only from the perspective of lesbian lives that women can imagine female sexualities that are not for just for others but *for* women, ourselves.

The perspective from lesbian lives also enables one to see the repressed lesbian story lines in real-life social relations and in literary texts that are purportedly about heterosexual relations. Readers can begin to see women's love for and erotic interest in other women in the lives of a Virginia Woolf or an Eleanor Roosevelt, and in the widespread romantic friendships recommended prior to the spread of Freudianism for what the age thought of as hetersexual women.[20] This perspective also leads to the identification of men's love for, erotic interest in, and total preoccupation and fascination with other men in most mainstream film and fiction and in intellectual projects in literature, philosophy, and other fields. One can begin to see the romance with the male proletarian characteristic of male intellectuals of the left; the continual construction of the exclusively male genealogies of influence that are characteristic of literary, philosophical, and political lineages; a certain omnipresent plot line in films and novels wherein two men indirectly relate "erotically" to each other through their competitive and even violent sharing of a woman.

An active female sexuality, a female sexual agency designed by and for women, would not have to be restricted to the bedroom, as is characteristic of white heterosexual women's sexuality in the bourgeois classes of the West today. This point can be understood from the perspective of many "other" women's lives, including lesbians'. Women's erotic energy would infuse their work, their public lives, their community relationships.[21] Men's involvement with their own "nonsexual" activity is often perceived as erotic—the scientist's with nature, the

19. Judith Roof made these points in conversation.
20. Faderman, *Surpassing the Love of Men.*
21. Audre Lorde, "Uses of the Erotic: The Erotic as Power," in *Sister Outsider: Essays and Speeches* (Trumansburg, N.Y.: Crossing Press, 1984).

artist's with his materials[22]—but heterosexual women are not permitted this infusion of their sexuality throughout their whole lives. As Freud understood clearly, the success of the process of becoming a (heterosexual) woman requires the restriction of libido, of sexual agency:

> I cannot help mentioning an impression that we are constantly receiving during analytic practice. A man of about thirty strikes us as a youthful, somewhat unformed individual, whom we expect to make powerful use of the possibilities for development opened up to him by analysis. A woman of the same age, however, often frightens us by her psychical rigidity and unchangeability. Her libido has taken up final positions and seems incapable of exchanging them for others. There are no paths open to further development; it is as though the whole process had already run its course and remains thenceforward insusceptible to influence—as though, indeed, the difficult development to femininity had exhausted the possibilities of the person concerned.[23]

Freud is writing of a historically specific type of woman: the type that showed up in his Vienna office during the late nineteenth century. For other women, their sexuality has certainly not been permitted to be reserved for the bedroom but expected to be available at all times—as a condition of work, or even in some cases of life—to any men in the dominant groups. This has been true, for example, of the sexuality of female slaves, peasants, household servants, and factory workers who could not claim a male "protector" nearby. These are other ways in which female sexual agency has been stolen from women. They are ways in which women have not been permitted or able to risk an active sexuality constructed both by and for themselves.

(5) A lesbian standpoint reveals the link between the oppression of women and the oppression of deviant sexualities. From the perspective of lesbian lives one can see the link between male supremacy and the oppression of "deviant" sexualities. In the first place, women's sexuality *is* the paradigm of deviant sexuality for traditional social and biological theorists. Aristotle claimed that women were inferior to men

22. Evelyn Fox Keller has discussed this point in *Reflections on Gender and Science* (New Haven, Conn.: Yale University Press, 1984).

23. Sigmund Freud, "Femininity," in *New Introductory Lectures on Psychoanalysis* (New York: Norton, 1963), reprinted in *Feminist Frameworks*, ed. Alison M. Jaggar and Paula S. Rothenberg (New York: McGraw-Hill, 1978), 98.

because their "semen" was "uncooked."[24] Others have seen female sexuality as an immature, poorly designed, or somehow lesser form of heterosexual male sexuality. The subjugation of women's sexuality is continuous with the subjugation of lesbian sexuality.

The repression of female sexuality is also continuous with the "perversions." Female sexuality is often conceptualized as animallike. In Western culture today, homosexual "man-boy" love is the object of criticism. But "man-girl" love (and even father incest) is virtually impossible even to see in everyday life or through the law. "Man-girl" love is the norm in reality and cultural images; even twenty- to fifty-year age differences between older men and their young wives are not uncommon (recollect Pablo Picasso, Pablo Casals, Justice William O. Douglas, Henry Kissinger). Anthropologists still say in public that incest is the fundamental cultural taboo recognized in every society, even though the statistics on child molestation by relatives they trust are horrendous. One is led to wonder whether "perversions" are yet another cultural object that men try to reserve for themselves, since apparently anything is supposed to be within the range of the normal for women (except, of course, refusing to make their sexuality available for men's use).

What is the best feminist politics around issues of sexual "deviance"? It is hard to say. On the one hand, feminism has wanted the state to step in and punish the perpetrators of rape, incest, and pornography that depicts violence to women for men's sexual pleasure. On the other hand, feminists are not the only ones to see that the social labeling and state regulation of what is proper and what is deviant sexuality is continuous with the social labeling and state regulation of female sexuality.[25]

Finally, feminism itself is often considered a perversity, as is illustrated by a remark attributed to Rebecca West: "I do not have the slightest idea what a feminist is, but I do know that I am called one

24. Aristotle, *De generatione animalium,* in *The Works of Aristotle,* ed. J. A. Smith and W. D. Ross (Oxford: Oxford University Press, 1908–52), 727a18, 766b20, 767b9, 775a15.

25. See Ann Snitow, Christine Stansell, and Sharon Thompson, eds., *Powers of Desire: The Politics of Sexuality* (New York: Monthly Review Press, 1983); Carole Vance, *Pleasure and Danger: Exploring Female Sexuality* (Boston: Routledge & Kegan Paul, 1984).

whenever I try to distinguish myself from a doormat." From the perspective of heterosexual lives, feminism is "unnatural," against "nature" (read: "patriarchal rule"), just as are homosexuality and any attitude toward sex and gender that male supremacy chooses not to legitimate. Feminists have often been labeled lesbian, since defenders of male supremacy cannot tell the difference between females wanting fully human rights (rather than only the rights women are permitted to have) and women refusing to devote their lives to men. In a culture where everyone is expected to be enthralled only by men and their achievements, such a confusion is understandable.

(6) A lesbian standpoint shows that gynephobia supports racism. Adrienne Rich has argued that gynephobia blocks white women's ability to identify with the concerns of women of color *as women* of color—with their concerns as mothers, daughters, economic providers, victims of sexual exploitation, and so forth.[26] Gynephobia used to make it easier for me to identify with the concerns of such men as Plato and Aristotle, whose daily lives were so different from mine, than with those of African American women in my classes or the African American woman who cleans my office. Racism, cultural differences, and, in many cases, class oppression ensure that we do not share the same experiences as mothers, daughters, or female workers. But gender, sex, culture, class, and two and a half millennia of history separate me from Plato's and Aristotle's lives. Thus, gynephobia hides the complicity of white women with racism. Women are supposed to hate each other and to compete for favor with men of the ruling groups. More generally, one can note that sexism, racism, class, and sexuality are used to construct one another on an everyday basis as well as in public policy (as I discussed in Chapters 7, 8, and 9). Intimacy commitments to men for whom their male supremacy is intertwined with their engagement in race and class struggles—whichever side of these battles they are on—makes it a race and class disloyalty for women to appreciate one anothers' lives across those race and class struggles.

(7) A lesbian standpoint suggests that the lesbian is a central figure in traditional masculine discourses. Paradoxically, the perspective from lesbian lives enables us to see that the lesbian is a repressed figure

26. Adrienne Rich, "Disloyal to Civilization: Feminism, Racism, Gynephobia," in *On Lies, Secrets, and Silence: Selected Prose 1966–78* (New York: Norton, 1979).

central to traditional male supremacist discourses: she is central in virtue of her absence.[27] How else can we explain the incredible force with which men and heterosexual women cannot see women's loving and caring relations to one another unless such relations are corralled into family relations? It is reasonable to suppose that this powerful nonseeing must be guarded by the threatening figure of a woman who loves, cares for, respects, is intrigued by, is devoted to, chooses to work or live with, or to have her imagination stirred by other women.

This brief discussion has done no more than suggest a few of the contributions to feminist thought that have already been made by starting thought in lesbian lives; no doubt there are others. What does this have to do with epistemology or with science?

A Lesbian Standpoint and Objectivity

Chapter 5 examined various differences between the lives of men of the dominant groups and of women—differences that make it valuable to start thought, to begin research, from the perspective of women's rather than men's lives. Less partial and distorted understandings of nature and social relations would result, I argued, if one began asking questions from the perspective of women's lives. Feminist research increases the objectivity of everyone's understandings by refusing loyalty to "the natives'" view of Western life and thought, where "the natives" are men in the dominant groups whose perspectives and interests have structured all our lives. Can analogous arguments be made about the scientific advantages to be gained for everyone by starting research from the perspective of lesbian lives?

Obviously so. First—to recapitulate the grounds for such a claim—lesbian lives have been devalued and neglected as origin points for scientific research and as the generators of evidence for or against knowledge claims. Second, if (heterosexual) women's exclusion from the design of social relations provides them with the valuable perspective of the stranger or outsider discussed in anthropology and sociology, lesbian exclusion can likewise be the source of new under-

27. Judith Roof, "Between Knowledge and Desire: Freud's Readings of Lesbian Sexuality," paper presented to Delaware Seminar in Women's Studies, University of Delaware, Newark, November 1989.

standings of the strange institution of compulsory heterosexuality that the indigenous peoples of the West have adopted. It can reveal as fundamentally distorted the ways of thinking about sex and gender that have flourished even within feminist writings.[28] Third, if the oppressed have fewer interests in ignorance about how nature and the social order actually work, then the perspective from lesbian lives will generate important new questions about, for example, how heterosexist control of sexuality supports capitalism, racism, and male supremacy.[29]

Fourth, the view from the perspective of lesbian lives is the view from the "other side" of sexuality struggles. Gaining knowledge is an active process, and political struggle is a great generator of insight—in the history of science no less than in other histories. The struggles that lesbians must engage in for survival can reveal regularities of social life and their underlying causal tendencies that are invisible from the perspective of heterosexual lives. Fifth, the everyday life of lesbians can reveal the caring for and valuing of women, the prioritizing of their welfare, the possibility of experiencing real intimacy and democratic domestic relations, which are only ambivalently enacted by men and heterosexual women.

Sixth, the perspective from lesbian lives permits various cultural "irrationalities" to emerge into clearer view: "normal" homosocial male worlds, but for women only heterosocial worlds—or homosocial ones "ruled" by men, the "normal" models of men's passions for and preoccupations with each other versus the "infantile" or "disorderly" characterization of analogous relations for women; "normal" female dependency and so-called male "autonomy," and so on. Seventh, many lesbians, and certainly lesbian intellectuals, are not just outsiders but "outsiders within."[30] The perspective from their lives, which are located not only on the margins of the social order but also in certain respects at its center, can reveal the causal relations between the margins and the center. Finally, one could construct sociological and historical arguments that this is the right time in history to begin thought from lesbian lives, to reflect on why changes in the social order make

28. See, e.g., Butler, *Gender Trouble*.

29. See, e.g., Michel Foucault, *A History of Sexuality*, vol. 1 (New York: Random House, 1980); Rubin, "The Traffic in Women."

30. Patricia Hill Collins, "Learning from the Outsider Within: The Sociological Significance of Black Feminist Thought," *Social Problems* 33 (1986).

possible the emergence of lesbians and gays as social "classes" not only *in* themselves but also *for* themselves.[31]

In the preceding section of this chapter I reviewed some of the substantive "scientific" insights to be gained by starting research from the perspective of lesbian lives. Here I have reviewed some *epistemological* grounds for arguing that doing so increases the objectivity of research. These are not the only epistemological grounds possible, nor is standpoint epistemology the only theory of knowledge one could use to justify the kinds of substantive claims reported above. For example, the feminist empiricist theory of knowledge outlined in Chapter 5 has its analogue in discussions of why research motivated by lesbian and gay social movements has been fruitful. There are no doubt other theories that justify the fruitfulness of such research in other ways. My effort here has been to show how feminist standpoints make it possible to ask new questions and see new things about nature and social relations not from the lives of a paradigmatic abstract woman (that is, heterosexual, white, Western, economically privileged) but from such specific women's lives as those of lesbians. As noted earlier, even the term "lesbian lives" is a cultural abstraction; race, class, sexuality, culture, and history construct different patterns of daily activity for lesbians as they do for the lives of others.

"Well," a critic might say, "perhaps starting inquiry from the perspective of lesbian lives is useful in the social sciences, but what possible difference could it make to the natural sciences?" There is no more (or less) reason to raise this question about a lesbian standpoint than about any other women's standpoint. I would go about answering this question in just the way I would proceed with respect to a similar question about "women's lives." This project lies ahead, but here are some preliminary questions.

What technologies are desirable from the perspective of the lives of women who do not live with men and who work primarily with each other? for women who do not live in conventional families? for communities where there is no division of labor by gender? These questions arise from the perspective of "single women's lives," but that formulation suggests a kind of impermanence and "lack" in their situation; it hints at fugitive supplies of handy jar openers and step stools for

---

31. Ferguson, *Blood at the Root;* D'Emilio and Freedman, *Intimate Matters.*

reaching high shelves. What if our housing, work, and transportation were designed to suit the lives of women who primarily work and live *with* other women and their dependents rather than only *for* men and their dependents? How would gynecology change if females bodies were no longer imagined to be fundamentally reproductive systems?

What would sciences look like that were no longer infused with subtexts and metaphors of active and autonomous women as sexually terrifying but, instead, with positive images of strong, independent women and with female eroticism woman-designed for women? These could not draw on sexist and misogynistic metaphors and models of nature or inquiry. They could not invoke as threatening the image of "wild and unruly" women or of women's energies devoted to women. They could not invoke the notion of dominating nature as a displacement for the domination of woman/mother. They *could* invoke metaphors of positive woman-to-woman relations—mother to daughter, colleague to colleague, pal to pal—in science's thinking about how nature is ordered and how the scientist should interact with nature. Would the prevalence of such alternative metaphors foster the growth of knowledge? If they were to excite people's imaginations in the way that rape, torture, and other misogynistic metaphors have apparently energized generations of male science enthusiasts, there is no doubt that thought would move in new and fruitful directions.[32]

What if white, Western women thought it more important that women in other races and classes be able to improve their lives than that white, Western men's exploitation of people in other races be supported? What views of science and new scientific and technological practices would emerge from such commitments? I take up this topic in the next chapter.

32. The role of metaphors in science is discussed esp. in Chapter 2.

II

Reinventing Ourselves as Other

More New Agents of History and Knowledge

Standpoint theories show how to move from *including* others' lives and thoughts in research and scholarly projects to *starting from* their lives to ask research questions, develop theoretical concepts, design research, collect data, and interpret findings. For feminist scientists and scholars, women have been the "Others" of special interest; the view from their lives could be used to generate research and scholarship that would provide less partial and distorted accounts of nature and social relations than had conventional work in the natural and social sciences, which claimed to start from no particular historical lives at all but simply to seek "the truth." With the few exceptions that prove the rule, it is only members of the dominant groups who are permitted the luxury of imagining that their lives and thoughts represent the ahistorically human. The rest of us are allowed to speak only as "women literary critics," "African American sociologists," or "the poor."

But women are to be found in every race, class, and culture. If feminist research and scholarship were to start from women's lives, they would have to start from *all* women's lives. Accounts of nature and social life from the perspectives of working-class and poor women, lesbians, and women of Third World descent in the First and Third Worlds have enriched and challenged the rest of feminist thought; that is, they have corrected the dominant accounts of the lives and thoughts of these groups of women at the periphery of North Atlantic societies. Just as important, however, they have challenged the dominant accounts of women's and men's lives at the "center." Chapters 8 and 9 identified some of the challenges to conventional accounts of the sciences which emerge if research starts not from the lives of people of European descent but from the lives of people of Third World descent. Chapter 10 pointed to some of the contributions to everyone's understanding of nonlesbian lives to be gained by starting from lesbian lives.

What is the relationship between experience and knowledge for the standpoint theories? I stressed earlier that what "grounds" feminist standpoint theory is not women's experiences but the view from women's lives. And in the preceding three chapters I argued that we can all learn about our own lives at the center of the social order if we start our thought from the perspective of lives at the margins. But these discussions still leave unclear just what relationship between experience and knowledge is being claimed.

There are two positions that the arguments of the preceding chapters have attempted to avoid. One is the conventional Western tendency to start thought from "the view from nowhere," to perform what Donna Haraway has called "the God trick."[1] This tendency, which might be called transcendental or ahistorical foundationalism, leads to parochialism because it can never recognize the possible *greater* legitimacy of views that claim to be historically situated but contradict the speaker's. The other is the tendency, in reaction to this ahistoricism, to insist that the spontaneous consciousness of individual experience provides a uniquely legitimating criterion for identifying preferable or less false beliefs. This can be thought of as experiential foundationalism— which obviously also tends to parochialism.

The challenge is to extract the illuminating kernel from its false and mystifying shell in each of these positions. On the one hand, we should be able to decide the validity of a knowledge claim apart from who speaks it; this is the desirable legacy from the conventional view. Otherwise, we end up committed to " might makes right" in the realm of epistemology, science, and the determination of "facts." (And the legacy is valuable even though the conventional view itself—through embedding this kernel in the chaff of ahistorical foundationalism— ends up supporting "might makes right" in science and knowledge-seeking.) On the other hand, it *does* make a difference who says what and when. When people speak from the opposite sides of power rela-

1. Donna Haraway, "Situated Knowledges: *The Science Question in Feminism* and the Privilege of Partial Perspective," *Feminist Studies* 14:3 (1988).

tions, the perspective from the lives of the less powerful can provide a more objective view than the perspective from the lives of the more powerful (Chapter 5 elaborated the rationale for this point). Moreover, until the less powerful raise their voices to articulate their experiences (frequently a dangerous act), none of us can find the perspective from their lives. For example, it is absurd to imagine that U.S. slaveowners' views of Africans' and African Americans' lives could outweigh in impartiality, disinterestedness, impersonality, and objectivity their slaves' views of their own and the slaveowners' lives. One doesn't have to assert that the slaves' views are impartial, disinterested, or impersonal in order to make this assessment, or to give the slaves—any more than one gives their masters—the last word about such matters as how the economics of slavery functioned on an international scale, in order to recognize the resource that the perspective from their lives' provides on the views typical of slaveowners.[2] If human knowledge is not in some complex way grounded in human lives and human experience, what is the source of its status as knowledge in modern Western societies? This notion is so central to the experimental spirit of modern science, so reasonably and widely accepted, as to look trivial on the page.

This chapter reflects further on the relationship that standpoint theory encourages between experience and knowledge. It examines the kinds of subjects or agents of history and knowledge and the kinds of projects that are generated by the "logic" of standpoint theory. I characterize as "reinventing ourselves as other" the standpoint enterprise that produces agents of history and knowledge who use experience in their knowledge-seeking in a different way from that of proponents of the two strategies to be avoided. That is, I explore here how to put into practice the "strong objectivity" criterion developed in Chapter 6.

Four Problems

The philosophical confusion about what relationship between experience and knowledge one should attempt to promote appears in four

---

2. This is why slave narratives are such a valuable source for historians. See, e.g., *Narrative of the Life of Frederick Douglass, An American Slave* (New York: New American Library, 1968).

ways in the lives of many of us: it is a scientific, epistemological, pedagogical, and political problem.

In scientific projects, ethnocentric assumptions distort the lives of marginalized peoples: racially marginalized people in the First and Third Worlds, the poor, sexual minorities, and women (to stay with the four most commonly discussed marginalizations). An explosion of new research and scholarship *by* these Others not only speaks about the conditions of their lives but also shows how ethnocentric learning distorts the lives of members of the centered groups: whites, the economically over-advantaged, the sexual " majority," and men. How can *we* actively study and learn about our dominant group selves and our culture without either replicating the conventional ethnocentric perspectives that rely on our spontaneous consciousness of the experiences in our lives, or inappropriately appropriating the experience of those Others whose voices have led us to see the need to rethink our views of ourselves? We have not *had* their experiences and do not live their lives.

Epistemologically, the standpoint theories argue that it is an advantage to base thought in the everyday lives of people in oppressed and excluded groups. But is it only the lives of the oppressed that can generate knowledge, especially liberatory knowledge? What can the role in knowledge-seeking be for the lives of those of us who are or would be white antiracists, male feminists, heterosexual antiheterosexists, economically overadvantaged people against class exploitation, and the like? And is articulated experience the grounds or the precondition for knowledge?

On the pedagogical front, teachers daily face the challenge to articulate or at least encourage a relationship between experience and knowledge-seeking which has not been preconceptualized. It is exciting for the Others in our classrooms today to see their lives and perspectives for the first time begin to be adequately reflected in the results of the research and scholarship generated by programs in women's studies, ethnic studies, African and African American studies, even occasionally in Marxist or "political economy" studies, and in gay and lesbian studies. In the United States many colleges and universities are beginning to revise their distribution and core curriculum requirements to include this research and scholarship in order to fulfill what they are coming to see as an important part of the mission of their institutions. But what about those students in our classes (or ourselves, for that matter) who apparently have the "wrong identities" or live the "wrong lives" for making contributions to this exciting new thinking? How can we encourage ourselves and our students and colleagues of European descent, men, heterosexuals, and the economically overadvantaged to think and act past the tendency to a guilt that is fundamentally inauthentic and passivity-inducing? We are not blank tablets, as the empiricists claimed; learning is an active process. How can we activate students and colleagues to function as subjects or agents of scientifically adequate, liberatory knowledge out of their own specific ("wrong") identities and social locations?

Finally, the political problem is how to encourage and energize the democratic tendencies and desires arising in social life. In our writing, teaching, and politics, how can we hail the most progressive tendencies in one another and in our students and colleagues? In particular, the question for those of us already engaged in liberatory efforts is how to create continuities and encourage progressive relationships between our projects. The old-style politics of unity—"either you're with us or you're against us"—is no longer effective or appropriate, if it ever was.

These issues are much too large and complex to be resolved here, but we can gain resources for further thought about them by pursuing the logic of the arguments in the preceding chapters.

Claiming Identities We Were Taught to Despise

It has been a struggle for feminists to learn precisely how to claim on behalf of our research, scholarship, and politics the perspectives that arise from our "despised" identities as women, as African American women, as women of color in the so-called First, Second, and Third Worlds, as lesbians, as poor and working-class women.[3] We have claimed the historical realities of our lives as the places from which our thought and politics not only do begin but also should begin. It has

3. My subhead is borrowed from Michelle Cliff's title: *Claiming an Identity They Taught Me to Despise* (Watertown, Mass.: Persephone Press, 1980). Much of the rest of this chapter will appear in an earlier version titled "Who Knows? Identities and Feminist Epistemology," in *(En)Gendering Knowledge: Feminists in Academe*, ed. Joan Hartman and Ellen Messer-Davidow © 1991 by the University of Tennessee Press. Hartman and Messer-Davidow's comments helped me to improve this chapter.

taken courage to claim these identities for such purposes when the gatekeepers of the intellectual traditions have insisted for centuries that we are exactly not the kinds of persons whose beliefs can ever be expected to achieve the status of knowledge. They still assert that only the impersonal, disinterested, socially anonymous representatives of human reason—a description that many of them explicitly intend to refer only to themselves—are capable of producing knowledge. Mere opinion is all that the rest of us can hope to produce. It is an extraordinary achievement, then, to have shown that the perspectives provided by our devalued identities can be epistemologically powerful, and that the unselfconscious perspective claiming universality is in fact not only partial but also distorting in ways that go beyond its partiality. The purportedly universal perspective is in fact local, historical, and subjective: only members of the ruling groups are permitted to elevate the perspectives from their lives to uniquely legitimate ones.

The political companion to this kind of feminist epistemological claim has been called "identity politics": working to emancipate one's group instead of "humanity." The group of African American women who wrote under the name Combahee River Collective explained: "Focusing upon our own oppression is embodied in the concept of identity politics. We believe that the most profound and potentially the most radical politics come directly out of our own identity, as opposed to working to end somebody else's oppression."[4] They were tired of hearing about how they should be concerned with liberating others, or how others were going to liberate them. We can call this position "situated politics" to emphasize what it shares with the "situated knowledge" discussed in earlier chapters: marginalized groups develop political and knowledge-seeking projects that originate explicitly from their own socially devalued lives instead of from "nowhere" or from somebody else's life.

The main question to be pursued here is this: Are there not additional forms of situated knowledge and politics hiding in the logic of these analyses? One can begin to detect other identities for knowers (for scientists, inquirers, thinkers, intellectuals, historical actors), sec-

---

4. Combahee River Collective, "A Black Feminist Statement," in *This Bridge Called My Back: Writings by Radical Women of Color,* ed. Cherrie Moraga and Gloria Anzaldua (Latham, N.Y.: Kitchen Table:Women of Color Press, 1983), 212.

ondary identities standing in the shadows behind the ones on which feminist and other liberatory thought has focused, identities that are also struggling to emerge as respected and legitimate producers of illuminating analyses. From the perspective of the fiercely fought struggles to claim legitimacy for the marginalized identities, these additional ones can appear to be monstrous: male feminists; whites against racism, colonialism, and imperialism; heterosexuals against heterosexism; economically overadvantaged people against class exploitation.

Should women spend time legitimating male feminist projects? Should African Americans devote energy to helping European Americans deal with their desire to be on the right side of race history, lesbians and gays to helping heterosexuals think through heterosexism, the poor to helping the rich deal with their exploitation of the poor? Isn't it the worst kind of agenda to force women back into their traditional role of collaborators, their conventional position as men's helpmates? Many female feminists would rather be run over by a truck. Moreover, it is clear that a lifetime commitment to such an agenda will ensure a female feminist a disillusioned and dreary life with plenty of intellectual and social deprivation. The "epistemological separatist" attitude that resists such projects has its valuable consequences, and I do not mean to suggest that every female feminist—certainly not most of them—should be expected to spend her time thus. Female feminists definitely should not be expected to do so by would-be male feminists. I do think, however, that feminism "for women only" is a luxury that female feminists can not afford and never desired. After all, we want to change the world—not only women.

Moreover, if feminism cannot legitimate male feminists and distinctive scientific and political projects for them, then how can feminists of European descent legitimately generate antiracist knowledge, academic feminists learn to see the world in ways informed by working-class women, heterosexual feminists learn to think from the perspectives of lesbian lives? What first appears external to the center of feminism—theorizing and and possibly helping to develop an agenda for male feminists—turns out, I am arguing, to belong at that very center. A feminism that is only of and for white women, for example, is nothing but a manifestation of self-interested individualism. It is what makes feminism part of other oppressed people's problem. But that stance immediately generates the Monster Problem: what does and should it mean to be a male feminist?

Contradictory Identities, Contradictory Social Locations

The standpoint theories themselves provide resources for answering such questions. From the perspective of conventional thought, all female feminists have identities that are contradictory; in other words, we think and act out of contradictory social locations.[5] Learning to think and act effectively out of those contradictions is an important part of claiming feminism, of becoming a feminist. A woman thinker is a contradiction in terms, as the feminist critics of science, philosophy, history, sociology, literature, and other fields have shown simply by quoting the statements of famous men about women and gender.[6] But in trying to specify why women and Western thought have a certain repulsion for each other, the critics have generated illuminating understandings of nature and social life today and in the past.

Bearing an identity or speaking from a social location that is perceived as a contradiction in terms can be a serious disadvantage within political, economic, and social structures, but such an identity can be turned into a scientific and epistemological advantage. In activating our identities as women scientists, women philosophers, African American women sociologists, lesbian literary critics—women *subjects and generators* of thought, not just objects of others' thoughts—we exploit the friction, the gap, the dissonance between multiple identities. Our identities appear to defy logic, for "who we are" is in at least two places at once: outside and within, margin and center. Learning to think from this "outsider within" social location has generated startling and valuable understandings in the social and natural sciences, literature and the arts. Black sociologist Patricia Hill Collins writes: "As outsiders within, Black feminist scholars may be one of many distinct groups of marginal intellectuals whose standpoints promise to enrich sociological discourse. Bringing this group, as well as those who share an outsider within status vis-à-vis sociology, into the center of analysis may reveal views of reality obscured by more orthodox approaches."[7] Chapters 5 and 6 suggested other ways to

5. "Identities" arise from "social locations" and, I shall argue, can also be chosen to legitimate social locations.

6. See Genevieve Lloyd, *The Man of Reason: "Male" and "Female" in Western Philosophy* (Minneapolis: University of Minnesota Press, 1984).

7. Patricia Hill Collins, "Learning from the Outsider Within: The Sociological Significance of Black Feminist Thought," *Social Problems* 33 (1986), S15. See also Bell Hooks, *Feminist Theory: From Margin to Center* (Boston: South End Press, 1983).

describe marginal intellectuals which bring out what it is about their social situation that generates scientific and epistemological "privilege" or "advantage." Being an outsider within captures central characteristics of these social locations.

The two (or more) identities of outsiders within are not just different from each other; they are oppositionally related. At the center one cannot speak at all *as* a woman, we are told, let alone as an African American woman, a lesbian, or a feminist. One can speak, uncomfortably for marginal people, only as a scientist, sociologist, literary critic, philosopher—the social shape, color, and historical activities of one's body to be not just not central in our speech but actively hidden, suppressed. It is somehow almost embarrassing to imagine a philosopher—one who is not a woman, not of color, or not a traveler from other cultures—mentioning in public as he begins his reflections on some typically "universal" philosophical issue that he speaks, of course, as a white male from the University of Delaware, for example. It would be embarrassing because from feminist perspectives it would be true that those details identified the location of his thought, but the speaker could not acknowledge why it was true and still conduct the discourse. A woman philosopher is a misfit because a man philosopher is an all too perfect fit. She is fundamentally impossible as long as he is so probable. Impossible people appear to us as situated in culture and history; it is their social situation that makes them impossible. Consequently, it requires no Sherlock Holmes to identify in a fairly precise way the cultural home of those with no putative social situation at all: they are at home in the world that has been designed as an extension of their desires and interests. "The view from nowhere" is generated by those who can afford the luxury of the "dream of everywhere."[8]

Feminist standpoint epistemology focuses on the scientific and epistemological importance of the gap between the understanding of the world available if one starts from lives of people in the exploited, oppressed, and dominated groups and the understanding provided by the dominant conceptual schemes.[9] In the feminist accounts, the para-

8. See, e.g., Susan Bordo, *The Flight to Objectivity* (Albany: State University of New York Press, 1987), for the history of this idea.

9. Recall that a standpoint is not a perspective; it takes science and politics to achieve a standpoint. Standpoints are socially mediated; perspectives are unmediated (see Chapters 5–7).

digmatic standpoint knowers are women who are intellectuals or researchers or wage workers or are active in the public world in some other way. These women are expected to fit their understandings of their own activities and of the rest of social life into Procrustean conceptual molds that were designed to explain to and for men in the ruling race and class the lives of people like themselves. It is out of this gap that feminist understandings have emerged.

But do these theorists intend the understandings generated by contradictory identities to be available, accessible, only to those women? Obviously not. For example, Patricia Hill Collins, quoted above, insists that people who do not have marginal identities can nevertheless learn from and learn to use the knowledge generated from the perspective of the outsiders within. Because Collins's contribution is to sociological discourse in general and to African American women, not only to marginal intellectuals such as herself, now we can all see aspects of race and gender relations that African American feminists first pointed out to us. Moreover, it is not just *about* African Americans that I learn *from* them; from Collins and other African American thinkers I have also learned certain things about European American experience, identity, and privilege which I previously took for granted simply as components either of human experience or of my purportedly individual experience. This line of reasoning leads to the understanding that feminist insights cannot be "for women only"; it cannot be that no one else can generate knowledge with the use of those insights, even though they are generated explicitly *for* women: that is, in order to explain the world for women instead of just for male supremacists.

Using the Contradictions

I find it paradoxical—and, frankly, suspicious—that most of the European-American feminists I know who admire, learn from, and use the understandings of feminists of color appear to overestimate their own ability to engage in antiracist thought but to underestimate men's ability to engage in feminist thought. They seem to believe that European American feminists are perfectly capable of generating antiracist analyses; recollect, they would say, the important contributions of Bettina Aptheker, Minnie Bruce Pratt, Adrienne Rich, and philosophers

277

such as Marilyn Frye, Margaret Simons, and Elizabeth V. Spelman.[10] Yet they also appear to believe that men cannot generate feminist analyses; forget, they imply, John Stuart Mill, Marx and Engels, and such recent writers as Isaac Balbus and Brian Easlea.[11] A kind of monster lurks in the logic of white feminist discourses: he is a white, economically privileged, Western, heterosexual man—and he is a feminist too. Does his existence discourage white, heterosexual, economically overadvantaged women from imagining that they should generate antiracist, antihomophobic, and anticlass analyses? Does such a monster also stalk through men's imaginations?

Some feminist epistemological positions have appeared to claim that only people who are women and therefore have women's experiences can generate feminist insights; that only those who are African American or lesbian or working-class or Third World can originate antiracist or antihomophobic or antibourgeois or antiimperialist insights. Is it true that *only* the oppressed can generate knowledge, that one can contribute to criticism and the growth of knowledge only out of one's own *oppression?* Standpoint theories argue that knowledge must be socially situated and that some situations are worse than others for generating knowledge. But is a "social situation" determined exactly and only by one's gender, or race, or class, or sexuality? How is the social situation of a woman who is a feminist scholar different from that of a woman who is an antifeminist, a misogynist, a defender of the

10. Bettina Aptheker, *Woman's Legacy: Essays on Race, Sex, and Class in American History* (Amherst: University of Massachusetts Press, 1982); Minnie Bruce Pratt, "Identity: Skin Blood Heart," in *Yours in Struggle,* ed. Elly Bulkin, Minnie Bruce Pratt, and Barbara Smith (Ithaca, N.Y.: Firebrand Books, 1988); Adrienne Rich, "Disloyal to Civilization: Feminism, Racism, Gynephobia," in *On Lies, Secrets, and Silence* (New York: Norton, 1979); Marilyn Frye, "On Being White: Toward a Feminist Understanding of Race and Race Supremacy," in *The Politics of Reality: Essays in Feminist Theory* (Trumansburg, N.Y.: Crossing Press, 1983); Margaret Simons, "Racism and Feminism: A Schism in the Sisterhood," *Feminist Studies* 5:2 (1979); Elizabeth V. Spelman, *Inessential Woman: Problems of Exclusion in Feminist Thought* (Boston: Beacon Press, 1988).

11. John Stuart Mill (and Harriet Taylor Mill), "The Subjection of Women," in *Essays on Sex Equality,* ed. Alice S. Rossi (Chicago: University of Chicago Press, 1970); Friedrich Engels, *The Origins of the Family, Private Property, and the State* (New York: International Publishers, 1942); Isaac Balbus, *Marxism and Domination* (Princeton, N.J.: Princeton University Press, 1982), chaps. 9–10; Brian Easlea, *Witch Hunting, Magic, and the New Philosophy* (Brighton, Eng.: Harvester Press, 1980); and Brian Easlea, *Fathering the Unthinkable* (London: Pluto Press, 1983).

patriarchy? What would it mean to say that it is the social situation provided by women's gender that makes the difference here?[12]

One can certainly identify reasons why female feminists are tempted to deny that there can be male feminists. They may focus on the fact that they don't agree with everything Mill, Marx, Engels, and others have said: such men were not *really* male feminists, because contemporary female feminists have criticized these men's views as not feminist enough. But where is the book or essay, feminist or not, by a woman or a man, with which anyone totally agrees? Moreover, each of us criticizes her own earlier views as not feminist enough now, or not feminist in the right way. Some female feminists may think it important to insist on a distinction between an antisexist and a feminist: men can be antisexists but not feminists. How people are named is a politically important matter, and I do not devalue the reasons to reserve the label "feminist" for females who work to improve women's lives. Nevertheless, it is important to keep in mind that feminists are made, not born. Biology is not enough to make Marabel Morgan or Margaret Thatcher feminists.

In other eras—the nineteenth century, for example—and especially in the context of biological determinist thought, distinctions between female and feminine and between feminine and feminist may have been virtually nonexistent, but we think only the ignorant confuse these terms today. Of course, we can *analyze* such phenomena as the way patriarchal culture constructs biology: that is, how gender constructs what counts as sex and, consequently, how sex relations are perceived and practiced.[13] And we can criticize the insistence on the sharp boundary between nature and culture of which the sex/gender distinction is part. But to engage in those analyses and criticisms is different from thinking it appropriate to use "female," "feminine," and "feminist" interchangeably, as do our students new to feminism. One must be able to distinguish biological, cultural, and political differences in order even to enter the recent discussions of the complex ways in which these differences have been and are used to construct one another.

Moreover, we have to note that feminists come in many different forms, and our agendas are often in direct conflict. Liberal feminism,

12. It is helpful to recollect Judith Butler's argument that gender is not fundamentally an interior state but a performative act; see *Gender Trouble: Feminism and the Subversion of Identity* (New York: Routledge, 1989).

13. Ibid.

Marxist feminism, radical feminism, socialist feminism, black feminism, the feminism of women of color, lesbian feminism, and Jewish,
Christian and Islamic feminisms are just a few of the feminist identities
and agendas appearing in recent book titles. Why, then, do critics of
the idea of feminist men reserve the label "feminist" for women with so
many different political agendas but refuse to bestow it on any man, no
matter what his political agenda or accomplishment?

Most likely the problem is far simpler. Men love appropriating,
directing, judging, and managing everything they can get their hands
on—especially the white, Western, heterosexual, and economically
overprivileged men with whom most feminist scholars and researchers
most often find themselves interacting in various workplace and social
institutions. Some have arrogantly tried to do so in the name of feminism, to claim a kind of feminist authority that as men they cannot
have, thereby inadvertently revealing that they have not grasped even
the most basic of feminist principles. They have tried to claim the name
for themselves without struggling against their own sexism—or, at
least, without enough struggle to earn them the minimum right to the
label.[14] Yet similar criticisms can be made about white people (including feminists) by people of color, about heterosexuals (including feminists) by lesbian and gay people, and about economically overadvantaged people (including feminists) by working-class people. In view of
these histories, one can understand why each marginalized group may
insist that the authority to say what is theoretically and politically
adequate research and scholarship must remain in the hands of the
marginalized. We have heard entirely too much from men about wom-
en and gender, from whites about people of color and race, from
heterosexuals about lesbians and gays and sexual preference, and from
economically overprivileged people about workers and the poor.
Claiming to be able to adopt the critical persona of the Other in the
name of her emancipation is unlikely to earn one the applause of the
Other. As Elaine Showalter pointed out, men who try to "read as
women" have tended to come off only as "Tootsies"—as men who

14. For illuminating discussions of these issues, see Susan Hardy Aiken, Karen Ander-
son, Myra Dinnerstein, Judy Lensink, and Patricia MacCorquodale, "Trying Transformations: Curriculum Integration and the Problem of Resistance," *Signs* 12:2 (1987);
Scarlet Friedman and Elizabeth Sarah, eds., *On the Problem of Men* (London: Women's
Press, 1982); Alice Jardine and Paul Smith, eds., *Men in Feminism* (New York: Methuen,
1987).

think they know more about women than women know and who refuse the difficult and painful work it takes to enter feminist discourse specifically as a man.[15]

It was easier in the "good old days" when only the speakers were "here" and the spoken-about were entirely "out there." Anthropologists have noted that their whole discipline has had to rethink its foundations and methods as the purported objects of its study migrated from the "bush," the plantation, and the "outback" into eminent authors' circles and into tenure and curriculum committees.[16] Men played an important and relatively unproblematic role as feminists in the nineteenth century, as theorist Stephen Heath points out in reflecting on whether or not he himself can be regarded as a feminist:

> Isn't John Stuart Mill's *The Subjection of Women* evidently a feminist book? No doubt. Here is a book written by a man that was clearly progressive and important for feminism, part of its history, Mill's intervention into the debate over the equality of the sexes widely recognized by women in the late nineteenth-century and subsequently as "an enormous advantage to the whole women's movement." No doubt. But *historically.* Today the history has changed, feminism has grown and advanced, there is no place in that way for a Mill.[17]

It would be interesting to explore just what should be counted as constituting the "growth and advance" that makes Heath think it more difficult for men to be feminists now than in earlier times. Presumably, one would have to include the recent feminist critiques of " normal" masculine sexuality; the discussions of the activities of the unconscious, of gender "codes," and of the construction of women as the Other; and certainly the ability of female feminists to gain access to most of the public forums once reserved for men. Women are now "here," not "out there"; we can speak for ourselves. Comparable advances would account for an analogous difficulty in the position now

---

15. Elaine Showalter, "Critical Cross-Dressing: Male Feminists and the Woman of the Year," in Jardine and Smith, *Men in Feminism*. Tootsie is the cross-dressing male character in the movie named for him, who states that he knows better than any woman could what it is like to be a woman. It is important to be able to criticize this stance on feminist grounds and still avoid "gay-bashing."

16. See, e.g., Clifford Geertz, "Being There" and "Being Here," in his *Works and Lives* (Stanford, Calif.: Stanford University Press, 1988).

17. Stephen Heath, "Male Feminism," in Smith and Jardine, *Men in Feminism*, 9. The quotation is from Millicent Garrett Fawcett, *Women's Suffrage: A Short History of a Great Movement* (London: Jack, 1912), 16.

of people of European descent in antiracist, antiimperialist, anti-colonialist movements.

Feminist standpoint theorists clearly intend insights generated from feminist standpoints to be used by men, though their silence on the issue is itself interesting. They may appear ambivalent about whether only women can generate accounts that are *for* women. But the logic of the standpoint argument leads to the assessment that oppression does not provide all the identities and social locations that can be used to ground the production of less false and distorted beliefs. On the one hand, standpoint theory asserts the importance of starting analysis from women's perspectives on our own lives, instead of from men's perspectives on either their own or women's lives. And it points to the importance of women's "bifurcated consciousness" for generating both the politics of the women's movement and the most radical feminist analyses. It is a struggle to articulate the forbidden, "incoherent" experience that makes possible new politics and subsequent analyses. On the other hand, this theory holds that our actual *experiences* often lead to distorted perspectives and understandings because a male supremacist social order arranges our lives in ways that hide their real nature and causes. Consequently, the politics of the women's movement is needed to bring women together as a group "for itself," and feminist analysis is needed to provide the tools to articulate our situations, which are occluded by the individual and untheorized experience each one of us has.

But if analysis and politics can clarify African American women's lives for African American women, can't that very same analysis, when conjoined with European American antiracist politics, also clarify European Americans' lives for them? And can't feminist analysis and feminist politics clarify men's lives for men? It is important to distinguish between "reading as a woman" and "reading as a feminist man," between seeing the world as a female and as a male feminist can. Deciding whether or not to bestow the term "feminist" on men is only part of the problem. At least equally important is the issue of how to specify in a positive way what the differences are between a feminist woman and a feminist man.

It would appear to be not a luxury but a necessity for feminism that European Americans should use the analyses provided by women (and men) of Third World descent to actively seek to understand European American lives. For me to do so is not an exercise in white narcissism, as some might think, but a necessary moment in understanding other people and my relations to them by understanding how I am situated in those relationships from the perspective of their lives. I am to enter this discourse precisely as a European American woman—not as purportedly colorless or as a white racist but as a white woman. I am to take responsibility for my identity, my racial social location, by learning how I am connected to other whites and to people of color; by learning what the consequences of my beliefs and behaviors as a European American woman will be.[18] The self-understanding I seek is to emerge as a result of my locating myself as a European American person in the analyses originally generated by thinkers of Third World descent and then continuing in the analyses by thinking about my world with the help of the accounts they have provided—yet still out of my own different social location. I can be only a white who intends to take responsibility for her racial location; I cannot be a person of Third World descent, seeking to take responsibility for that different social location. As noted earlier, the validity of our claims must be largely independent of who makes them, but it still matters who says what, when, and to whom.

Is it not also a necessity to encourage men—to demand of them—that they start *their* thought by locating themselves as members of the group called "men" who are the objects of discussion in feminist analyses, and then to continue in that direction by thinking about themselves and the world around them with the help of the feminist analyses—yet with a creative focus on and from *their* lives, which, after all, only they "have"? Of course, we women all watch them having those lives; we hear what they say about them; we wonder at the peculiar paths between the events and environments we see and the strange things they say about those events and environments and ponder why they behave in the ways they do. One's life is not entirely private; it is not a secret possession that dooms one to solipsism.

One more consideration: the Combahee River Collective authors say that "[we] find it difficult to separate race from class from sex oppression because in our lives they are most often experienced simultaneously."[19] Yes, it is clearly a luxury of being in the dominant race

18. Bettina Aptheker discusses the importance of thus taking responsibility for one's identity in *Tapestries of Life: Women's Work, Women's Consciousness, and the Meaning of Daily Life* (Amherst: University of Massachusetts Press, 1989).

19. Combahee River Collective, "Black Feminist Statement," 213.

and class that allows some women to experience their oppression only "as a woman." As European American intellectuals we are not oppressed by race hierarchy or, most of us, even by class structures. But we can *learn* to experience the race and class relations in which we participate. We can learn the causal connections between the events in our lives and those in the lives of people of color and of poor people. We can learn to experience our race and class situation as one that gives us race and class overprivilege. I can learn to experience the male supremacy that shapes my life as precisely the kind of male supremacy to which women in my class and race will be subjected, and as different from the kind of male supremacy to which an African American or a poor woman will be subjected. We can learn to find it difficult to separate race and class overadvantage from the particular forms that male supremacy takes in our lives. If it is a virtue for the women in the Combahee River Collective not to deny the race and class dimensions of the socially constructed experience within which they apprehend the world, why is it not also for the rest of us? We cannot learn and generate less distorting explanations and understandings out of an identity and a social location that are half or two-thirds repressed. If we take the stance that we can separate or ignore these dimensions of our lives, how are we different from the men in the dominant social groups who claim that they can separate the authority of *their* knowledge claims from the social situations that generated their claims?

But if we as white women can become knowers in this way, then so can The Monster. If he can't, then neither can we. His destiny appears to be linked to ours.

Three Surprising Consequences of Standpoint Logic

We are now in a position to note that the feminist standpoint logic has three outcomes that appear surprising from the perspective of the logic generally attributed to the Marxist theory that provided its epistemological resources. First, the subject of feminist knowledge—the agent of these less partial and distorted descriptions and explanations—must be multiple and even contradictory. This is true in two different senses. For one thing, women exist only in historically specific cultural forms—rich or poor, black or white, Italian or Chinese, heterosexual or lesbian. Cultural forms include also being a maker versus a consumer of scientific knowledge, and a daughter versus also a mother. These various lives are in many respects in conflict—not just different but opposed—yet each is potentially a starting point for feminist knowledge, for generating less partial and distorted accounts of nature and social life. Thus, feminist thought or knowledge is not just one unitary and coherent "speech" but multiple and frequently contradictory knowings. As the subject of feminist knowledge, the class "women" is multiple.

Further, each individual feminist knower is also multiple in a way that mirrors the situation of women as a class. It is the thinker whose consciousness is bifurcated, the outsider within, the marginal person now also located at the center, the person committed to two agendas that are themselves at least partially in conflict (socialist feminist, black feminist, Jewish feminist, woman scientist), who has generated feminist sciences and new knowledge. It is thinking from a contradictory position that generates feminist knowledge. So the logic of the standpoint directive to start thought from women's lives requires starting from multiple lives that are in many ways in conflict with one another and each of which has its own multiple and contradictory commitments. (In contrast, the subject of knowledge for both the conventional empiricist philosophy and for Marxism was supposed to be unitary and coherent.)

Consequently, and second, the logic of standpoint theory requires that the subject of liberatory feminist knowledge must also be the subject of every other liberatory knowledge project. Since lesbian, poor, and black women are all women, feminism will have to grasp how gender, race, class, and sexuality are used to construct one another. It will have to do so if feminism is to be emancipatory for marginalized women but also if it is to be maximally scientific for dominant-group women about their own situation. There would otherwise be no way to distinguish between feminism and the narrow self-interest of dominant-group women—just as conventional androcentric thought permits no criterion for distinguishing between "best beliefs" and those that serve the self-interest of men as men.

This is also what other liberatory movements must do to accomplish *their* goals. That is, worker movements must look at their worlds from the perspective of women workers' lives too; antiracist movements must look at their issues from the perspective of the lives of women of color too; gay liberation movements must start their thought from lesbian lives too. What feminist thought must know must inform the thought of every other liberatory movement. Nor is it only the women in those other movements who are to "know" the world from the perspective of women's lives; everyone in those movements must do so if the movements are to succeed in reaching their own goals. This requires that women be active directors of the agendas of the movements. But it also requires that men in those movements be able to generate original knowledge about themselves and the world from the perspective of women's lives—as John Stuart Mill, Marx and Engels, and later male thinkers have done.

Third, therefore, women cannot be the unique generators of feminist knowledge. Women cannot claim this ability to be uniquely theirs, and men must not be permitted to refuse to try to produce fully feminist analyses on the grounds that they are not women. Men too must contribute distinctive forms of specifically feminist knowledge from and about their particular social situation. Men's thought too will begin from women's lives in all the ways that feminist theory, with its rich and contradictory tendencies, has helped us all—women as well as men—understand how to do. It will start there in order to gain the maximally objective theoretical frameworks within which men can begin to describe and explain their own lives in less partial and distorted ways. This is necessary if men are to generate anything more than "gender nativism" or "gender folk belief" about themselves and the world they live in. Historians argue that the past can illuminate the present by creating a broader context and a contrasting perspective from which to examine present-day institutions and social practices. For the same reasons, the "distant present" can illuminate the "near present"; for men, the view from women's lives can illuminate their own lives by creating a broader context and contrasting perspective from which to examine critically the institutions and practices within which occur their own beliefs and behaviors, social relations and institutions. Men must struggle to create for themselves a kind of experience of their own gender location which male supremacy has forbidden.

Having women's experiences—being a woman—clearly is not sufficient to generate feminist knowledge; all women have women's experiences, but only at certain historical moments do any of us ever produce feminist knowledge. Our experience lies to us, and the experiences of the dominant gender, class, race, and sexuality produce more airtight, comprehensive, widely believed, and tenacious lies. Dominant-group experience generates the "common sense" of the age that is such a bizarre and fascinating subject matter to anthropologists and historians. All of us must live in social relations that naturalize, or make appear intuitive, social arrantements that are in fact optional; they have been created and made to appear natural by the power of the dominant groups. Thus, it is not necessary to have any *particular* form of human experience in order to learn how to generate less partial and distorted belief from the perspective of women's lives. It is "only" necessary to learn how to overcome—to get a critical, objective perspective on—the "spontaneous consciousness" created by thought that begins in one's dominant social location. I say "only" ironically, because many women have found the process of understanding their lives through the lenses of feminist theory to be an extremely painful process, even though an energizing one that has given them, they say, a "second birth." Obviously, it is also a painful process for men who find themselves abandoning cherished beliefs about themselves and their worlds and choosing new behaviors and life projects following their "second birth."[20]

These three consequences of feminist standpoint logic have their analogues for each of the other liberatory movements. For example, the subject or agent of African American knowledge is multiple; it must be the subject or agent of every other emancipatory knowledge also, and therefore African American knowledge cannot be generated only by African Americans. Let me follow through briefly the logic of this apparently arrogant claim. If African Americans exist only in historically specific configurations of class, gender, sexuality, and culture, then the subject or agent of African American philosophy or sociology or science must be multiple. The lives that provide the starting points for African American thought will then also be providing the starting points for feminist, socialist, gay and lesbian, and other emancipatory thought. They are part of the multiple subject or agent of *every* emancipatory thought. Thus it is not only African Americans who must have the obligation to generate knowledge from the perspective of African American lives. A critique by a white Western philosopher

20. Mary O'Brien, *The Politics of Reproduction* (Boston: Routledge & Kegan Paul, 1981), has argued that making (male supremacist) culture and history is men's "second birth." If so, making feminist culture can be their "third birth."

which starts off from analyses made by African Americans and others of Third World descent may not be called "African American philosophy"—it is a matter of great political importance how thought is named—but it must *be* African American thought if it is to become a maximally objective account of Western philosophy.[21]

"Traitorous" Identities and Social Locations

The kind of contradictory identity and social location that is a consequence of following standpoint logic in fact has already appeared in a number of analyses. Chapter 10 presented some substantive contributions to feminist thought in general which were made by starting from lesbian lives: for example, that heterosexual women and men and gay men, not just lesbians, can now think about the possibility of women's choosing intimacy *and* autonomy rather than having to choose between them; that there are homoerotic subtexts in films, novels, and intellectual traditions (and in our own everyday and institutional social relations) which on the surface appear heterosexual. With the assistance of analyses that began in lesbian lives (and were almost always first generated by lesbians), lesbians and heterosexuals have learned to "read against the grain" their otherwise spontaneously heterosexist experience (heterosexist assumptions have shaped lesbian

21. Contrary to the way some may choose to interpret this argument, it provides reasons to attract and hire more African American philosophers and others of Third World descent, to create an increase in their numbers instead of the decline we face at this historical moment. American philosophy is impoverished by its small number of practitioners of Third World descent both in leading "white" graduate institutions and in the historically black colleges. They, not philosophers of European descent, have reported how the world looks from the perspective of African American and other Third World lives, and Western philosophy looks very parochial indeed from these perspectives: see, e.g., Leonard Harris, ed., *Philosophy Born of Struggle: Anthology of Afro-American Philosophy from 1917* (Dubuque, Iowa: Kendall/Hunt, 1983); V. Y. Mudimbe, *The Invention of Africa: Gnosis, Philosophy, and the Order of Knowledge* (Bloomington: Indiana University Press, 1988); "Philosophy and the Black Experience," special issue, *Philosophical Forum* 9:2–3 (1977–78); and the work of Cornel West, e.g., *Prophecy Deliverance: An Afro-American Revolutionary Christianity* (Philadelphia: Westminster Press, 1982), esp. chaps. 1–3. Similarly, it is no accident that it has been primarily women who have developed feminist perspectives. Thought reflects the social situation of those who produce it, but life "situations" are not biologically determined: people who find themselves in politically regressive social situations can and must make changes if they would change their politics and thought.

and gay "experience" too, as historians point out). Some people whose sexual identity was not "marginal" (in the sense that they were heterosexual) have "become marginal"—not by giving up their heterosexuality but by giving up the spontaneous consciousness created by their heterosexual experience in a heterosexist world. These people do not think "as lesbians," for they are not lesbian. But they do think as heterosexual persons who have learned from lesbian analyses.

Similarly, whites *as whites* can provide "traitorous" readings of the racial assumptions in texts—literature, history, science—written by whites. For example, in a widely discussed essay, historian Minnie Bruce Pratt activates her identity as white and as lesbian in the service of antiracist understanding of her own cultural inheritance. She reenvisions the southern town in which she grew up, contrasting the partial and distorted vision that her father, loyal to the Confederacy, tried to give her with what she knows of it after relearning history from the perspective of black lives. She tries to come to terms with the fact that the beautiful tree on the edge of town—the one under which she had so often picnicked as a child—was one from which many a "strange fruit" had hung, in the famous phrase from Billie Holliday's song about lynching. She struggles to fit together her identity as a white, southern woman who is lesbian—to find positive uses for all of that identity—and her anger and hatred of what the South stands for to racist southerners.[22] White lesbian poet Adrienne Rich has made a similar effort.[23]

Though these feminists speak out of their own social locations, they speak not out of their own racial oppression (for they are not racially oppressed) or unanalyzed white supremacist assumptions but out of a critical reflection on their own world as antiracists who are white. They are trying to know, first-hand, what that world is and how one sees it through a critically *reflective,* but still white, consciousness. Their contributions to how *we* see the world are distinctive exactly because of the ways they have found to activate their full identities and social situations as whites and to let us see how they are taking respon-

22. Pratt, "Identity." See also discussions of Pratt's analysis in Karen Kaplan, "Deterritorializations: The Rewriting of Home and Exile in Western Feminist Discourse," *Cultural Critique* 6 (1987); Biddy Martin and Chandra Talpade Mohanty, "Feminist Politics: What's Home Got to Do with It?" in *Feminist Studies/Critical Studies,* ed. Teresa de Lauretis (Bloomington: Indiana University Press, 1986).
23. See Rich, "Disloyal to Civilization."

sibility for their identity as whites, how they are discovering the causal connections between their own social situation as whites and the situations of blacks in the past and today.

A puzzling question: is it significant that although they have not experienced white supremacy as blacks do, they *have* experienced oppression as women and as lesbians? Does one have to have had the experience of some oppression or other in order to generate the kinds of traitorous analyses I am discussing? Could something as politically innocent as the experience of oppression under "adult supremacy" be enough? Can we "recollect" experiences we have never had, just as we remember the fear or horror or confusion or pleasure we felt as we identified with the lot of fictional or historical characters—Oedipus or Antigone, Frederick Douglass or Sojourner Truth? Or can one take responsibility for one's social identity without ever having had an experience of oppression?

Many white men do not have to seek the experience; they are Jewish, or poor, or in some other way an "oppressed minority." But I suggest that we should refuse to believe that there are no ways for overly privileged white men too to learn to take responsibility for their identities. Most of them may not *want* to do this, but they *can*. Some have. We can ask, where is that great, powerful, autonomous will that supposedly only people like them possess—the will we've had to hear so much about in Western philosophy? Are we going to let them get away with the claim that *they* are more bound by their historical identity and location than they routinely expect us to be in their classes on Aristotle, Hobbes, Rousseau, and the other "dead white boys" in whose preoccupations they expect us to find it reasonable to structure our thought?

It is clear that intellectual and political *activity* are required in using another's insights to generate one's own analyses. I cannot just repeat, robotlike, what African American thinkers say and never take responsibility for my own analyses of the world that I, a European American, can see through the lens of their insights. If I did so, I would be thought of as stupid, or as insidiously devaluing their thought and undermining the legitimacy of its analyses. If, when asked about a race issue, I never venture beyond statements such as "Patricia Hill Collins says . . . ," I thereby imply that I am reserving my own opinion on the matter.[24] It is important to give credit where credit is due, and European Americans

---

24. Collins, "Learning from the Outsider Within."

frequently fail to do so when speaking on race issues. But I must learn how to see the world differently for myself in an active and creative way through the theoretical and political lenses that African American thinkers originally constructed to produce distinctive insights. A functioning antiracist—one who can pass "competency tests" as an antiracist—must be an actively thinking antiracist, not just a white robot "programmed" to repeat what blacks say.

The possible birth of a similar kind of traitorous male feminist agent of thought and action is the topic of struggle for many of the authors, male and female, in the recent collection of work by literary critics titled *Men in Feminism*.[25] Those writers, women and men, ask of men that they undertake the difficult, painful, but important project of analyzing themselves, their lives, and their worlds within the terms of the analyses that have emerged from female feminists: "You have everything to tell us about your sexuality, your relations to your mother, to your fathers" (both biological fathers and "fathers" in the intellectual traditions), and also "to death, scopophilia, fetishism, . . . the penis and balls, erection, ejaculation (not to mention the phallus), madness, paranoia, homosexuality, blood, tactile pleasure, pleasure in general, desire, . . . voyeurism, etc."[26] This is not to be a repetition of the misogynous reports on such topics characteristic of "normal" men's speech, of Freud's "normal male misogyny," but a report informed by feminist theory and practice. As male literary critic Stephen Heath puts the point: "We should probably start by trying to grasp *who we are as men,* asking that from feminism rather than wondering what 'they' want from an assumed male us."[27] These men are not to try to "read as a woman," or to tell female feminists how to think, or to commit the multitude of other sins that men exhibit in the name of their purported feminism when they fail to distinguish between what it is in the power of women and of men in feminism to do (though female feminists have not been entirely clear on this difference either). Instead, they are to speak specifically *as men* of themselves, of their bodies and lives, of texts and of politics, using feminist insights to see the world as men who are as knowledgeable about female-generated feminist analyses as female feminists expect one another to be. Nobody will do this

25. Jardine and Smith, *Men in Feminism.*
26. Alice Jardine, "Odor di Uomo or Compagnons de Route?" in ibid., 60. Jardine attributes the first request to Hélène Cixous.
27. Heath, "Male Feminism."

perfectly, but the point is to develop strategies that encourage men as well as women, whites as well as people of color, straights as well as gays and lesbians, the economically overadvantaged as well as the working class and the poor to become active agents of historical understanding.

Generating *new* insights through research and scholarship originating in traitorous social locations, however, is not the only or perhaps even the most important task that people who choose such locations can undertake. Men who want to be part of the solution rather than only the problem for feminism can promote the understandings produced by women feminists. They can teach and write about women's thought, writings, accomplishments. They can acknowledge their debts to feminism. They can sponsor women students and feminist projects for men students. They can criticize the sexism and androcentrism of their male colleagues and of male thinkers, writers, and public figures. They can move material resources to women and to feminists. They can insist on redesigning practices and policies in workplaces, communities, and families in ways that will close the gap between the social resources available to women and to men. In short, they can be politically active as feminists. After all, as men they have access to economic, political, social, and psychological resources that are often still not available to women. They can show themselves and the world that they really are "in feminism" and not just using feminism to advance their careers or their stature in the eyes of their female friends and colleagues.[28] As female feminists know, political struggle is a precondition for knowledge: men *will* discover what patriarchal power is really about as they candidly criticize their male colleagues' sexism to those colleagues.

But if female feminists expect men to undertake such agendas, then we feminists (women and men) must engage in these same struggles as whites against racism, as heterosexuals against heterosexism, as economically overprivileged people against class oppression, as Westerners against imperialism (insofar as each of us fits these categories). Let us develop traitorous agendas on behalf of these other projects so that we may learn better how to be "disloyal to civilization" in all the ways it so richly deserves.

I hope it is clear that my intention is to make it both harder and easier to *become* a male feminist, a white antiracist, and so forth. For

28. See Jardine, "Odor di Uomo," 60–61, for a similar list.

example, I must undertake difficult tasks in order to generate effective antiracist insights. As I said, I cannot just repeat what people of color have said. I have to educate myself about people of color, their struggles, and their cultures. I have to study my own ignorance as well—the culturally rewarded white ignorance discussed by philosopher Marilyn Frye.[29] I have to study white exploitation, domination, oppression, and privilege. I have to generate the kinds of explanations of these conditions that I expect men to generate of the conditions that give them privilege. This is to be a competency-based antiracism, a competency-based male feminism. If these processes are not painful, I am probably not doing them right. After all, it can't be entirely a pleasure to discover the unintentionally racist assumptions that have guided so many of my thoughts and practices—especially at those moments when I was exactly trying to enact a piece of antiracist business. So achieving a traitorous identity or social location requires the performance of difficult and painful tasks.

Some people enjoy the challenge of such tasks. Articulating the requirements for achieving traitorous identities provides them with real agendas. Some people would rather learn difficult truths about themselves and their world than suspect that they are thinking and behaving disreputably. (Most of us would probably prefer to learn these truths gently. We don't always get what we prefer, but we can console ourselves with the recollection that if we belong to a privileged group, we are already destined to get much more of what we prefer than the oppressed are of getting what they prefer.) The current prevalence of mixed messages to whites from people of color and to men from women—"do as I do, but do not dare to do as I do"—has the effect of immobilizing and pacifying the very people who do want to develop disloyalties to "civilization." Specifying agendas for achieving traitorous identities and social locations makes it easier to achieve them. We have no more excuses for complaining that we're wrong if we don't and wrong if we do.

Another good reason for developing traitorous social locations is that if I cannot learn to think critically out of traitorous identities, my ways of seeing race and class will tend to focus on the oppression of others rather than on my own situation and the perspective available from within it. It is persons of my kind of race and class, after all, who perpetrate racism and class exploitation. If I fail to activate a traitorous

29. Frye, "On Being White."

perception of my race and class, then implicitly I encourage the kind of "studying down" that feminists have criticized in sexists. Furthermore, in learning about my race *only* from racists or, quite differently, *only* from people of color, I deprive myself of *my* perspective on myself. I fail to take an active role in defining what my racial identity can be. I accept the passive, nonresistant, noncreative role assigned to me by others. I permit myself to be inserted as a robot, as an object, into their dramas—though I am directed to play quite different roles in dramas authored by the dominant ideologies and in those authored by oppressed peoples. (And I do not mean to suggest that only "bad" roles are imagined for me by people of color and the economically disadvantaged.) There have always been people of European descent who struggled publicly and in their personal lives to undermine racism and imperialism, to use their white overprivilege to undermine its institutionalization. They have used the view from Third World lives (generated originally by people of Third World descent) to gain a less partial and distorted understanding of their own situation. This, too, is what "being white" can mean to whites.

Finally, many of us want to recruit women and men to feminism and other counterculture studies and politics. But people are not enthusiastic about participating in such efforts if they are constantly told that they are the wrong kind of people to speak in this group and that consequently their learning can only be passive—rehearsing what others have thought up for them to think. We need to devise agendas of activity for all the social identities and social locations that our potential recruits carry. If women, people of color, gays and lesbians, and the economically disadvantaged can create counterculture agendas for themselves, then so can men, whites, heterosexuals, and the economically overadvantaged.

The relationship between experience and knowledge is a no less difficult issue for the liberatory knowledge movements than it has been for the conventional Western discourses. The liberatory movements cannot escape addressing experience's contradictory character as, on the one hand, a highly prized and difficult to obtain precondition for the possibility of creating knowledge from the perspective of the lives of marginalized peoples and, on the other hand, as the residence of "obvious" but highly distorting culturewide beliefs. The preceding reflections have attempted to identify some resources for moving past the

Scylla of "the view from nowhere" and the Charybdis of experiential foundationalism. They appropriate objectivity not only for the views from marginalized lives, provided by the people who live those lives, but also for the uses that can be made of that theoretical framework by people who choose to *become* "marginalized." Those of us in the overly privileged groups cannot succeed in giving up the privilege that the social order insists on awarding us. Men will be perceived to be and treated as men, no matter how feminist their behavior; whites will be perceived to be and treated as whites, no matter how antiracist their agendas. It is crucial to avoid imagining that men and people of European descent in the dominant groups really do lead marginal lives in the ways that women and people of Third World descent are forced to do. But we can learn to think and act *not* out of the "spontaneous consciousness" of the social locations that history has bestowed on us but out of the traitorous ones we choose with the assistance of critical social theories generated by the emancipatory movements. We can note that this stance enacts at least some of the agendas postmodernists have called for.

Pedagogically, this approach provides resources for energizing our students and colleagues to seek objective perspectives on their own lives as a way to produce liberatory knowledge out of their own (transformed) social situations. And politically, we can try—respectfully, tentatively, but firmly—to create continuities between our various liberatory social movements and to encourage progressive relationships between them. The point is not that women should "bond" with individual men, people of Third World descent with individual whites, and so on, but rather that our scientific and political movements should establish grounds for "solidarity." Reasoning out possibly viable agendas is not a full solution to the challenge to produce such solidarities, but it can contribute resources to this project.

The knowledge-seeking and political projects of liberatory movements have reinvented the dominant cultures as alien and bizarre traditions from which they can learn techniques and against which they can define their own agendas.[30] What better alternative could there be for moving toward more democratic societies than for those of us in the overly favored groups to join them and reinvent ourselves as "other?"

---

30. This is V. Y. Mudimbe's way of stating the project for African philosophers today (*The Invention of Africa,* 171).

12

Conclusion

What Is Feminist Science?

In one form or another and with various concerns in mind, the question "Can there be feminist science?" has been raised by virtually everyone who participates in or contemplates the feminist discussions of the sciences.[1] This chapter examines the reasons for the negative or only limited positive responses of some feminist writers and then considers how the arguments of the earlier chapters make it reasonable to think that there *already are* specifically feminist sciences and that it is

1. See, e.g., Lynda Birke, *Women, Feminism, and Biology* (New York: Methuen, 1986); Ruth Bleier, *Science and Gender: A Critique of Biology and Its Theories on Women* (New York: Pergamon Press, 1984), and her Introduction to her edited collection, *Feminist Approaches to Science* (New York, Pergamon Press, 1986); Anne Fausto-Sterling, *Myths of Gender: Biological Theories about Women and Men* (New York: Basic Books, 1985); Elizabeth Fee, "Women's Nature and Scientific Objectivity," in *Woman's Nature: Rationalizations of Inequality,* ed. Marian Lowe and Ruth Hubbard (New York: Pergamon Press, 1983), and her "Critiques of Modern Science: The Relationship of Feminism to Other Radical Epistemologies," in Bleier, *Feminist Approaches to Science;* Donna Haraway, "Primatology Is Politics by Other Means," in Bleier, *Feminist Approaches to Science;* Evelyn Fox Keller, "Woman Scientists and Feminist Critics of Science," *Daedalus* 116:4 (1987): 77–97; Helen Longino, "Can There Be a Feminist Science?" in *Feminism and Science,* ed. Nancy Tuana (Bloomington: Indiana University Press, 1989); October 29th Group, "The October 29th Group: Defining a Feminist Science," *Women's Studies International Forum* 12:3 (1989); Hilary Rose, "Hand, Brain, and Heart: A Feminist Epistemology for the Natural Sciences," *Signs* 9:1 (1983), and her "Nothing Less than Half the Labs," in *Agenda for Higher Education,* ed. Janet Finch and Michael Rustin (New York: Penguin Books, 1986); Sue Rosser, *Teaching Science and Health from a Feminist Perspective* (New York: Pergamon Press, 1986), and her *Feminism in the Science and Health Care Professions* (New York: Pergamon Press, 1988).

beneficial to both feminism and the sciences to continue developing them.

The Diversity of Answers

One must expect multiple answers to any question about "feminism" and "science" because, as I noted at the outset, the two terms are themselves contested zones. In some subcultures in the West and more extensively in the Second and Third Worlds, the term "feminist" is an epithet used against women who defend women's interests publicly or their own interests in more intimate settings. Sometimes it is used to place distance between the speaker and bourgeois, racist, Eurocentric, and heterosexist tendencies in feminism. And in the United States and elsewhere, one frequently finds women or men who insist that they are not feminists but who vigorously advocate agendas that are indistinguishable from those that have been advanced as specifically feminist. These people have equated "feminism" with the more radical positions that have been taken within feminist theory and politics.

Among feminists who claim the label, groups with varied and conflicting agendas compete to define what should be counted as feminism. Liberal feminism, socialist feminism, African American feminism, Third World feminism, lesbian feminism, and postmodernist feminism could all articulate distinctive standards for alternatives to the dominant preoccupations of the sciences that we have. Moreover, feminists work in or experience the sciences in different ways. The constraints within which workers in physics, U.S. health care, technology transfers to the Third World, and sociology address the issue of imagining feminist science are dissimilar in many respects.

Science critics, including those who are feminists, think about science within the agendas of different metastudies of science: traditions in the philosophy of science and sociology of science can encourage conflicting answers to the question of whether there can be feminist science. Furthermore, exactly what is and is not to be legitimately regarded as science is contested within the natural and social sciences and the philosophy and social studies of science, as well as in contemporary public debates about such issues as creationism, holistic medicine, behaviorism, and sociobiology. Thus, both feminism and science have been constructed through continuing struggles for political resources and will remain contested terrains as long as they are thought to provide such resources. Echoes of these disputes can be heard in feminist reflections on how to think about a positive program for the sciences.

Before proceeding, I must emphasize that in considering the question of *"feminist"* science," I do not intend to discuss either *female* or *feminine* science—whatever those might be. Some critics have chosen to interpret feminist discussions of science as advocating something they refer to as "female science." At times they appear to think that feminist science discussions are critical of the influences of male biology and therefore are advocating the influence of female biology on science—a position I have never seen articulated anywhere but in these antifeminist writings. At other times it appears that in eliding the differences between female, feminine, and feminist, either these antifeminists cannot distinguish between sex, gender, and a theoretical-political position, or they are attempting to discredit feminist meta-science discussions by implying that such discussions are concerned only with absurd issues. Whatever these critics may have in mind, the issue here is about the possibility of *feminist,* not female, science.

Other critics of feminism—and a few feminists—have conflated feminist science and something they refer to as *feminine* science.[2] It is true that feminism reevaluates "the feminine"; all the feminist thinkers who have pondered the possibility of feminist science have rejected tendencies in the dominant culture to devalue and ignore the distinctive insights and skills that women can bring to social life, including to science. So a reevaluation of the activities assigned to women, of the skills and talents developed in performing those activities, and of what can be learned by starting thought from women's lives is an important aspect of feminist thought in every disciplinary area. And since science and femininity have been coconstructed again and again in opposition to each other in the very same political arenas, it is tempting to try to disrupt this pattern by refusing, or trying to bridge,

2. Proponents of "gynocentric science" are sometimes concerned with "feminine" science in the sense of sciences that incorporate what are thought to be distinctively feminine modes of interacting with or thinking about nature; at other times, the term "gynocentric science" refers to women's scientific practices that have not been recognized as science; occasionally it is used to mean feminist science.

the gap between the two.[3] But most feminists have been critical of many aspects of femininity as well as of masculinity, because both genders are socially constructed within gender relations designed and maintained by the dominant institutions. Women have been systematically excluded from leadership positions in these institutions and thus from the design of the gender they must bear.[4] As a result, most feminist thinkers (and probably also most of those who are uncritically enthusiastic about femininity) would strenuously object to identifying feminist science with feminine science.

To return to the question: one useful way to conceptualize the diversity of existing feminist responses to the possibility of feminist science is to distinguish three groups of thinkers who give a negative or only limited positive answer to the question and a fourth group comprising those who respond positively but are concerned only with the social sciences and, possibly, biology.

Only "After the Revolution"

First, there are those who think that although significant advances toward feminist sciences can be made now, the term "feminist science" should be reserved for practices and institutions in the future.[5] Large-scale social changes will have to occur, they believe, before anything develops that could reasonably be thought of as feminist science.

Ruth Bleier has summarized feminist characterizations of the feminist science of the future, pointing out that any such list must remain provisional at this early stage. To begin with, scientists must acknowledge that their values and beliefs influence their scientific practices and learn to identify the effects. They "would have to be explicit about their

3. See, e.g., Londa Schiebinger, *The Mind Has No Sex: Women in the Origins of Modern Science* (Cambridge, Mass.: Harvard University Press, 1989), for the history of the science/femininity dichotomy.

4. See Judith Butler, *Gender Trouble: Feminism and the Subversion of Identity* (New York: Routledge, 1990), for the argument that gender is not fundamentally an interior state but a performative act repeated daily.

5. Good examples of this argument are provided in Birke, *Women, Feminism, and Biology;* Bleier, *Feminist Approaches to Science;* Fee, "Women's Nature and Scientific Objectivity"; and October 29th Group, "The October 29th Group." This appears also to be Longino's position (in "Can There Be a Feminist Science?"), although she argues that the term "feminist science" should mean "doing science as a feminist."

assumptions; honest, thoughtful, and careful in their methods; open in their interpretations of each study and its significance; clear in describing the possible pitfalls in the work and their conclusions about it; and responsible in the language used to convey their results to the scientific and nonscientific public." Moreover, feminist approaches to science would "aim to eliminate research that leads to the exploitation and destruction of nature, the destruction of the human race and other species, and that justifies the oppression of people because of race, gender, class, sexuality, or nationality." Feminist science

> would not be elite and authoritarian and, therefore, it would have to be accessible—physically and intellectually—to anyone interested. It would be humble and acknowledge that each new "truth" is partial; that is, incomplete as well as culture-bound. Recognizing that different people have different experiences, cultures, and identifications (therefore, different perspectives, values, goals, and viewpoints), feminist science would aim for cultural diversity among its participants, so that through our diverse approaches we would light different facets of the realities we attempt to understand. Such diversity would help to ensure sensitivity of the scientific community to the range of consequences of its work and thus its responsibility for the goals of science and the applications and by-products of its research.

Many of these changes cannot occur, Bleier points out, until scientists

> reconceptualize science, its methods, theories, and goals, without the language and metaphors of control and domination. And for *this* to happen to a significant degree, profound changes must occur in a system that is based on power, control, and domination and that recognizes and rewards those who support and reinforce its ideologies and aims. However, just as destroying all nuclear weapons and outlawing their manufacture is not equivalent to world peace, but nonetheless an urgent first step, so can feminists continue to take our first steps toward the transformation of the ideologies and practices of the modern institution of science.[6]

Bleier's point about metaphors has been discussed by many feminists. Carolyn Merchant, Evelyn Fox Keller, and other historians of science have noted that reliance on metaphors and models of both nature and research processes which center forms of order and relationships that are idealized in bourgeois, Western notions of mas-

6. Bleier, *Feminist Approaches to Science*, 15–16.

culinity results in partial and distorted descriptions and explanations of nature and social life.[7] Metaphors and models that stress context rather than isolated traits and behaviors, interactive rather than linear relations, and democratic rather than authoritarian models of order in both research and nature may enable less partial and distorted descriptions and explanations. These alternative metaphors and models have been associated with womanliness; their use can be thought of as infusing values arising from women's lives into scientific practices and outcomes.

Historian of science Elizabeth Fee outlines reasons for reserving the term "feminist science" for a possible future science:

> At this historical moment, what we are developing is not a feminist science, but a feminist critique of existing science. It follows from what has been said about the relationship of science to society that we can expect a sexist society to develop a sexist science; equally, we can expect a feminist society to develop a feminist science. For us to imagine a feminist science in a feminist society is rather like asking a medieval peasant to imagine the theory of genetics or the production of a space capsule; our images are, at best, likely to be sketchy and insubstantial. There is no way of imagining, in advance, a fully articulated scientific theory. We are, however, free to play with ideas and to consider the criteria that a feminist science should fulfill, but we should not confuse this with the actual production of scientific theory, nor should we take our inability to imagine a fully developed feminist science as evidence that a feminist science is itself impossible.[8]

A few feminists in laboratories here or there are in no position to redesign the programs of the natural sciences. Agendas for the extraordinarily expensive research in physics, chemistry, and much of biology are set in international councils and are heavily dependent on state and industrial funding.

These are persuasive arguments for reserving the term "feminist science" for the empirical knowledge-seeking that will be the successful consequence of feminist and other liberatory struggles to transform the societies in which science occurs. Their proponents insist that changes in the laboratories and science councils can be only partial

7. Evelyn Fox Keller, *Reflections on Gender and Science* (New Haven, Conn.: Yale University Press, 1985); Carolyn Merchant, *The Death of Nature: Women, Ecology, and the Scientific Revolution* (New York: Harper & Row, 1980).
8. Fee, "Women's Nature and Scientific Objectivity," 22.

and ineffectual unless liberatory movements can also change the social order for which science is such an important resource. And they lead us to think strategically about just what changes in science and in society would most effectively advance that goal.

Nevertheless, the argument of this book leads to the different conclusion that feminist sciences are already being developed. If we extract the notion of scientific knowledge-seeking from its Eurocentric, natural science, and positivist paradigms—an enterprise for which the preceding chapters have attempted to provide resources—we can continue to pay close attention to the foregoing concerns and still find good reasons to conceptualize this broader kind of feminist attention to science as developing specifically feminist sciences. But I am getting ahead of the argument of this chapter.

Only Good or Bad Science

Some observers have argued that what feminists want to refer to as feminist science is really just plain good science. For one thing, they say, many of the changes feminists argue for have been proposed by others at various times in the history of science. If nonfeminists today or in the past have also called for the changes feminists call for, then these changes are not specifically feminist, and there is no need to talk about such a thing as "feminist science." Well-known biologist and historian of science Stephen Jay Gould has made this argument.[9] So have members of an organization of feminist natural scientists, the October 29th Group, who report: "When the question of whether such a thing as 'feminist science' could exist was first posed, our training in traditional science made it seem absurd: good science is good science whatever your politics, we thought."[10] (Some members of this group subsequently changed their minds; others did not.)

However, the conventional dichotomy between "good science" and "bad science" is not the whole story. As biologist Anne Fausto-Sterling points out, it is no accident that it is feminists—most but not all of whom are women—who have called for changes resulting in what

9. See, e.g., Stephen Jay Gould's review of Bleier's *Science and Gender, New York Times Book Review,* August 12, 1984, p. 7 (cited in Fausto-Sterling, *Myths of Gender,* 208).
10. October 29th Group, "The October 29th Group," 254.

Gould and the October 29th Group refer to as "good science." These are people "who have waged intense battles for the opportunity to do scientific work in the first place. Their very status as outsiders—women and feminists in a masculine scientific world—has lent them a vision which quite appropriately claims the label of feminist." Fausto-Sterling explains that the revisions feminists have called for, "although they may represent good science, arose in the context of a vast and multiply branched political-cultural movement, that of modern Western feminism. To hold out for a good versus bad science analysis is to ignore the important role feminism has played in *forcing* the re-evaluation of inadequate and often oppressive models of women's health and behavior." Thus, there is real danger that good science will not happen if we do not recognize the role of political change in the advance of knowledge.

> In the past, legions of highly trained doctors and scientists have failed to see and criticize what is wrong with the biomedical and behavioral models of female behavior. Why? Because, I believe, they had no alternate framework within which to develop new sight. Feminism provided that new vision, allowing many scientists—even those who do not consider themselves political feminists—to move in a new direction. "Good science" in the absence of a political and cultural movement did not get very far. . . . Quality research alone is not enough. *Good science*—which in this historical moment incorporates many insights from feminism—can prevail only when the social and political atmosphere offers it space to grow and develop.[11]

Stephen Jay Gould has in fact come close to this understanding when writing in another context about advances in the history of science:

> Science, since people must do it, is a socially embedded activity. It progresses by hunch, vision, and intuition. Much of its change through time does not record a closer approach to absolute truth, but the alteration of cultural contexts that influence it so strongly. Facts are not pure and unsullied bits of information; culture also influences what we see and how we see it. Theories, moreover, are not inexorable inductions from facts. The most creative theories are often imaginative visions imposed upon facts; the source of imagination is also strongly cultural.[12]

---

11. Fausto-Sterling, *Myths of Gender*, 209, 213.
12. Stephen Jay Gould, *The Mismeasure of Man* (New York: Norton, 1981), 21–22.

The insistence that there can be only good science or bad science but not feminist science leads in confusing and counter-intuitive directions. Should we say that the misogynist and racist sciences that Gould criticizes were good science as long as no scientists knew about (listened to? took seriously?) criticisms of them? This interpretation would justify the refusal of scientists to seek out or even to listen to criticism. Alternatively, should we say that because all science must be open to future revision, no one ever does or in principle could produce good science? These logical extensions of the good science/bad science dichotomy reveal the problems with that conception of scientific practice.

Misguided Efforts

Comments made here and there in other contexts by poststructuralist feminists suggest another answer. It is important, they say, to disrupt dominant discussions and practices but to refuse to provide positive accounts of "what we should do" or "how the world really is." Criticisms are valuable in themselves, whether or not one can imagine viable alternatives. They set people thinking in new directions. They make space for the emergence of new thoughts and actions that are too fragile to be forced into comprehensively elaborated doctrines or plans of action. The immediate move to alternative agendas and accounts can prematurely truncate critical discussion. The demand for alternative agendas as a condition of finding reasonable any preceding critique can encourage simplistic or simply wrongheaded solutions to problems that are still only partially grasped. Moreover, isn't it arrogant, as well as a delusion, for one individual or a particular group to presume to be able to state "how the world is" or "what we should do"? The rush to "practical" solutions refuses the hard work, both intellectual and political, necessary to arrive at them. It can have the consequence of ensuring only impractical ones.

The poststructuralist stance raises important issues. Nevertheless, one must recognize that it is unlikely to satisfy policymakers, who need to know what the best assessments are today of "how the world is" and "what we should do about it" as they attempt to remedy and forestall the damage done to women's lives by sexual violence, the health-care system, the court system, dangerous technologies, and a profit-oriented economy. It is a great strength of the first two positions described that in attempting to envision feminist sciences they engage existing natural and social science institutions. Even the apparent conservativeness of attempting only to correct "bad science," or to create a feminist sociology or biology that simply is " good sociology" or "good biology," must be balanced against the radical consequences of vigorous and continuing attempts to move from "here" to "there"—to transform the institutions within which scientific knowledge-seeking now occurs. For most women in the natural and social sciences, the possibility of beginning to transform those institutions is the condition of their continued work in them. The rest of us need them there (as argued in Chapter 3), for how could we have feminist sciences in the future without any women scientists?

Feminist Social Sciences Are Sciences

Those who argue that there can be feminist sciences but not yet, who hesitate to countenance the possibility of feminist sciences here and now, appear to have only the natural sciences in mind. In the social sciences and even in biology and medicine, however, where disciplines contain a multiplicity of research traditions, the idea of creating now a feminist sociology, economics, biology, anthropology, or psychology does not seem so far-fetched. In these areas feminist efforts are competing with existing nonfeminist research agendas in the kinds of contests for explanatory success and institutional legitimation that were common before feminism arrived on the scene. African American sociology, Marxist anthropology, holistic medicine, and sociobiology are familiar to us as research traditions that compete within a plurality of theoretical approaches in their respective fields. In these contexts, feminist sociology, or anthropology, or medicine, or biology can be defined simply as scientific knowledge-seeking that is directed by existing feminist theories and agendas.

The Need for Disloyalty to Conventional Assumptions

All the tensions and disputes about a positive scientific program for feminism are fruitful. Feminist knowledge-seeking will have to meet standards for adequacy arising from the many different needs of women and men to create sciences that can provide adequate descriptions and explanations of nature and social relations. I suggest, however, that several assumptions or positions in much of this discourse are excessively loyal to conventional beliefs that feminists otherwise criticize. One such position follows Anglo-American conventions in restricting the term "science" in its central or paradigmatic meaning to the natural sciences, in contrast to the European practice of seeing as equally deserving of the label "scientific knowledge-seeking" those modes of systematic and effective inquiry that are favored in the social sciences and even in humanities and arts. Another regressive assumption holds that insofar as the feminist program for the sciences is still disputing the foundations of this field, it cannot yet be said to have achieved the status of a science. Critical discourse about foundations is a symptom of a pre-science, according to this view; only when "normal" research programs are in place and no one is any longer debating how to do feminist science will feminism have got itself a real science.[13] A related position is excessively constrained by remnants of positivist agendas in that it does not conceptualize metatheories of science as an integral part of science proper. It assumes the necessity of a sharp separation between a theory about how to gain empirical knowledge and the actual gaining of empirical knowledge: only the latter is to be thought of as really science (see Chapter 3). Finally, a Eurocentric stance, that "real science" is only what the modern West has done or chooses to call science, falsely assumes that it makes no sense to imagine effective sciences with very different theories of nature, different approaches to gaining empirical knowledge, and different agendas from those of modern Western science. This assumption has the (inadvertent) consequence that the sciences of the Chinese and African high cultures (see Chapter 9) may not appropriately be called sciences (though I doubt that the feminists whose views I have summarized here would want to be committed to such a position).

Expanding Feminist Sciences

As a contribution to the ongoing debate, I want to suggest another answer that draws from some of the central themes of this book. The claim that feminist sciences already exist articulates understandings of

---

13. Thomas Kuhn developed this conservative position in *The Structure of Scientific Revolutions* (Chicago: University of Chicago Press, 1970).

feminism and of science that are parallel to ideas emerging in other emancipatory science movements and in the new social studies of science. Because the most comprehensive feminist sciences come into view at the intersection of the transformed logics of science and of feminism explored in earlier chapters, let us review the trajectories of these two transformations.

The Transformed Logic of Science

Consider, first, what feminist science would be if our conception of science were sufficiently comprehensive to enable the sciences to examine critically in a systematic way all the evidence for their claims—or, at least, significantly more evidence than they are now competent to recognize. Chapter 6 discussed standards for a notion of "strong objectivity" that would be competent to identify and critically examine types of evidence that now go largely unexamined within the sciences: namely, beliefs and assumptions shared by a narrowly conceived scientific community. Feminism needs sciences that are more objective than the knowledge-seeking practices of androcentric, bourgeois groups in the West which have been passed off as objective, dispassionate, disinterested, universal science. Women—and men—cannnot understand or explain the world we live in or the real choices we have as long as the sciences describe and explain the world primarily from the perspectives of the lives of the dominant groups. If we must learn about this society and its nature only from its "natives"—that is, within conceptual frameworks and agendas restricted by the needs and desires of the peoples indigenous to the ruling class, race, and gender—we cannot gain objective explanations of nature, or of either our lives or theirs. The conventional sciences in the West are in these respects only "folk science"; that is, they are highly complex systems of belief that are effective for achieving all sorts of goals that these "natives" have. Sometimes some of us "non-natives" benefit from those projects, although we have a better chance to do so if we share the gender, class, or race of the "natives." But the successes of the sciences and their technologies frequently are achieved only at great cost to the other races, classes, and gender whose labor and suffering have made possible these benefits for the few. These kinds of problems with what the "natives" refer to as science have been amply detailed in earlier chapters and by other critics.

The elite politics of Western science is supported by conceptual inadequacies. We are now in a position to appreciate more fully four features of the sciences' social relations with nature that the conventional views of science do not conceptualize. First, we must say of science that it is politics by other means but also that it can produce reliable empirical information; it can do so as, for better or worse, it participates in politics. Second, science contains both progressive and regressive tendencies, and it leaves itself open to manipulation by regressive social forces to the extent that its institutions do not acknowledge and grapple with these contradictory internal features. Third, the observer and the observed are on the same causal plane, shaped by the same kinds of social forces. Thus in crucial respects nature-as-an-object-of-scientific-knowledge simulates culture in that it can appear to us only as already socially constituted, and it is socially reconstituted through scientific processes, among others. Moreover, the same social forces that shape nature-as-an-object-of-knowledge and other parts of culture also shape us and our scientific accounts—the best as well as the worst. Fourth, all sciences do assume epistemologies (theories of what knowledge is, what makes it possible, and how to get it), and both sciences and epistemologies implicitly hold sociologies of knowledge (theories about the social characteristics of the production of knowledge). But disciplinary traditions and conventions in philosophy, the social studies of science, and the sciences themselves lead practitioners in these fields to hold primarily archaic, excessively narrow, or suppressed or unconscious epistemologies and sociologies of knowledge.[14]

This transformed logic of the sciences decisively breaks with assumptions about the autonomy of the natural sciences from both society and from social theory. The fact that the natural sciences do not take humans or their social relations—historical intentionality—as the objects of their study does not exempt them from needing social theories and research agendas as central components. Certainly, much of the world that the natural sciences try to describe and explain is at the outermost limits of intentional life: that is, it has none. (As various forms of intentionality are discovered to extend further and further into the animal world, the borders between the natural and the social themselves shift.[15] But they have not shifted enough to invalidate this

14. These problems are discussed esp. in Chapters 4 and 7.
15. For a startling discussion of just how far they have shifted into the animal world, see Barbara Noske, *Humans and Other Animals: Beyond the Boundaries of Anthropology* (Winchester, Mass.: Pluto Press/Unwin Hyman, 1989).

claim.) The "oddity" of a lack of intentionality in much of the natural world, however, does not exempt the natural sciences from the directives of critical social science to examine the constraints and resources placed on belief formation by different historically located social relations. The natural sciences must incorporate the critical, self-reflexive methods beginning to emerge in the social sciences in order to block the intuitive, spontaneous consciousness of nature and inquiry to which all of us, but especially scientists, are susceptible.[16] The natural sciences, I have argued, are a particular kind of social science and should be so conceptualized. Only in this way can a strong objectivity be activated, one that insists on socially situated science and on scientific rather than "folk" accounts of those social situations.

Given this conception of science, why hesitate to say that feminist sciences exist now? They consist of the feminist metatheories of science and the research programs in the natural and social sciences that these metatheories already direct. In the social sciences and biology such theories and programs are already in place, even though most of the work in those fields is still not organized in these ways. But the theories and research practices generated by feminists are increasing in status— not decreasing. And after all, although some people still find useful the astrological and other now discredited practices that were central to scientific knowledge-seeking in the premodern and early modern West, these have been pushed to the periphery of modern Western scientific knowledge-seeking. Is it not reasonable, then, to imagine a similar fate awaiting research projects in sociology that purport to explain community organization or the structure of occupations without considering feminist criticisms of conventional approaches to these topics? or research projects in economics that set out to explain patterns in poverty or in international "development" without considering feminist criticisms of conventional approaches?

It is true that in areas of science such as physics and chemistry, there just aren't any research programs directed by feminist theories about science. There are feminist criticisms of various aspects of physics and chemistry, analyses of the role of the natural sciences in promoting androcentrism, and the beginnings of less sexist social relations in the labs. But research in these fields is organized on an international level, and thus is less susceptible to the kind of feminist influences that have been effective in more locally organized social research projects. These

16. See Roy Bhaskar, *Reclaiming Reality* (New York: Verso, 1989), chap. 4.

disciplines may never develop feminist or any other kind of knowledge-seeking agendas that advance liberatory social needs—or they may. At this stage of their development it is at least a possibility that they will remain captured by the elite politics of their patrons[17]—unless those who find such a future alarming take action to ensure that it does not happen. But the emergence of feminist science depends not on whether *all* fields of contemporary science can be transformed but only on whether some processes of seeking knowledge about the natural and social worlds can be developed which are directed by feminist rather than androcentric goals. As new sciences emerge, others become marginalized. Alchemy and astrology have been left behind as useful to the advance of the science of their day (according to historians of science) but not relevant to the needs of the later generations who continued to develop Western sciences. Recollecting historians' reports of the intimate relations between the development of modern science and the development of warmaking in the early modern period,[18] one can hope that a postmodern world will not need many of the sciences that the modern West has found so useful.

A Transformed Logic of Feminism

I can more quickly address the transformed logic of feminism because it was the topic of the last chapter. I argued there that the subject/agent of feminist knowledge is multiple and sometimes even contradictory in that women are located in every class, race, sexuality, culture, and society. Feminist thought consequently is multiple and even contradictory as it starts from the lives of all these different kinds of women. Moreover, the subject/agent of feminist knowledge is also the subject/agent of every other liberatory knowledge project: the liberation of women workers must be central to the agenda of workers' movements, of lesbians to gay liberation movements, of women of Third World descent to Third world liberation movements, and so forth. Those movements need feminist knowledge in order to achieve their own goals, just as feminism needs *their* knowledge.

17. This issue was discussed in earlier chapters. See Paul Forman, "Behind Quantum Electronics: National Security as the Basis for Physical Research in the U.S., 1940–1960," *Historical Studies in Physical and Biological Sciences* 18 (1987).

18. See, e.g., Margaret Jacob, *The Cultural Meaning of the Scientific Revolution* (New York: Knopf, 1988), 251.

Therefore, in a certain sense, women are not uniquely the agents of feminist knowledge. Feminist thought must ground its critical examination of nature and social relations in women's lives. But men too must learn how to do this from their own particular historical social situations as men who are traitors to male supremacy and, more generally, to conventional gender relations. And analogous consequences of the standpoint logics direct the knowledge projects of the other liberatory movements.

The issue of the relationship between experience and knowledge is still problematic in the liberatory epistemologies. It is important to preserve the integrity and wisdom inherent in culturally specific experiences, and especially in previously ignored and devalued ones, while also coming to terms with the fact that experience in many respects hides the realities of our lives: experience "lies." It is necessary to avoid the "view from nowhere" stance of conventional Western epistemology while refusing to embrace the exaltation of the spontaneous consciousness of our experiences—the experiential foundationalism—that has too often appeared to be the only alternative. One reasonable position is to say that the experiences of more powerful groups lie more often, more deeply, and more tenaciously than the experiences of the less powerful, though the latter must also be scientifically, causally explained. The possibility of generating knowledge of the lives of the more powerful from the perspective of the lives of the less powerful, however, is not to be *given* to well-intentioned members of more powerful groups; instead, those of us who have such "privileges" of power and intend to use them to increase social justice must *demand* it of ourselves as what our professed good intentions require of us.

Finally, it must be noted that it is not just the logic of the standpoint *theories* that leads to the claims made for them. If a theory "forced" one to assent to politically distasteful, depressing, and counterintuitive claims, then one could regard those consequences as in themselves good reasons to find the theory implausible. Instead, the extended logic of the standpoint theory explains certain patterns of belief and behavior that are already well developed in contemporary intellectual and political life. Effective progressive political workers and thinkers are increasingly informing themselves about what the politics and analyses of other emancipatory movements demand in terms of how such workers and thinkers should understand their own lives and thoughts, though the clearest statements of the necessity and requirements for such new sciences have appeared only in feminist writings so far.

The feminisms of "feminist science" are already moving toward more scientific feminisms, in the sense that they are becoming "strongly reflexive" (see Chapter 6). They are beginning to grasp the historical location of their own analyses and the consequences of these analyses for the lives of others. Thinking from women's lives provides crucial resources for the reinvention of sciences for the many to replace sciences that are often only for the elite few. Without such new sciences, privileged groups remain deeply ignorant of important regularities and underlying causal tendencies in nature and social relations, and of their own location in the social and natural world. Without such sciences, the majority of the world's peoples remain deprived of knowledge that could enable them to gain democratic control over the conditions of their lives.

Whose sciences? Whose knowledge? The answers to these questions are up to us.

Index

Abstract masculinity, 118, 131, 158
Affirmative action, 51–53, 191, 205
Africa:
  European underdevelopment of, 227–31
  history of science in, 223–27
African American feminism, 7, 131, 176, 275, 277. *See also* Women of color
African Americans, 198–99, 287–88
Amin, Samir, 153, 177
Androcentrism:
  epistemologies and, 47–49, 111, 116–17
  research bias and, 39–42, 111, 116–17
Aptheker, Bettina, 129, 176, 254, 257, 277
Ayer, A. J., 165

Bacon, Francis, 43, 64, 147–48
Bad science, critique of:
  androcentric bias and, 111
  vs. critique of science-as-usual, 54–56
  features of, 57–58
  feminist contributions to science and, 61–67, 74–75
  feminist science as good science and, 302–4
Balbus, Isaac, 278
Beauvoir, Simone de, 257

Belenky, Mary, 118, 122
Bhaskar, Roy, 79, 170
Biology, feminist research in, 108, 112
Black women. *See* African American feminism; Women of color
Bleier, Ruth, 299–300
Bloor, David, 168
Boch, Gisela, 179
Bunch, Charlotte, 256
Butler, Judith, 13

Causal asymmetry, 166–68
Class:
  critique of science-as-usual and, 67
  emergence of system, 132–33
  feminism as problem and, 193
  integration with gender and race, 212–17
  opportunities for women and, 22–23
  racial discrimination and, 205, 216
  women in sciences and, 200–201
Collins, Patricia Hill, 124, 131, 176, 275, 277, 290
Combahee River Collective, 273, 283
Competency standard, 217, 292–93, 305–6. *See also* Strong objectivity
Consequences of science:
  responsibility for, 2–4, 14–16, 37–38, 88–93
  Western story of science and, 219–20